PORPHYRA

Harvesting Gold from the Sea

PORPHYRA

Harvesting Gold from the Sea

Ira A. Levine
Dinabandhu Sahoo

I.K. International Publishing House Pvt. Ltd.
NEW DELHI • BANGALORE

Published by
I.K. International Publishing House Pvt. Ltd.
S-25, Green Park Extension
Uphaar Cinema Market
New Delhi – 110 016 (India)
E-mail: info@ikinternational.com

ISBN 978-93-80026-94-7

Published by Krishan Makhijani for I.K. International Publishing House Pvt. Ltd., S-25, Green Park Extension, Uphaar Cinema Market, New Delhi – 110 016 and Printed by Rekha Printers Pvt. Ltd., Okhla Industrial Area, Phase II, New Delhi – 110 020.

Preface

The intent of this book is to offer a primer to any individual, family, village cooperative, NGO, state, federal funding agency, or corporation that wishes to introduce *Porphyra* (known as Nori in Japan) farming as a socio-economic development driving force. This effort is not intended to be a textbook on *Porphyra* biology but a reference tool for nori cultivation from the conchocelis culture system through blade harvest and primary processing. Our hope is to provide aquaculture extension agents a teaching guide for nori cultivation. This guide will provide step-by-step instruction for the assembly of a nori production system, care, maintenance, harvesting, and product preservation. The nori training course is one step in the support system necessary to recruit, train, and assist the formation of individual and cooperative nori growing concerns.

Farming of seaweeds in general and nori in particular represent a sustainable alternative to coastal subsistent fishing villages; increasing disposable income for the fisherman while decreasing the ecological damage that certain fishing practices exert on local environments. The historical organization of the family farm, village cooperative, or emerging corporate nori operations all represent both ecological and sustainable opportunities. With proper site selection, nori farming represents no threat to coral reef ecosystems.

The introduction and organization of nori farming has many examples to build upon. There are countless examples of well-established nori farming cooperatives, based on long-standing fishery cooperatives or village organized efforts. Unlike seaweed farming for phycocolloid extraction, "buying stations" are replaced by nori auctions or direct sales to nori marketing companies. Prices are a function of annual harvest yields and sheet quality. Farmer's income varies significantly as a function of the farm size, quality of cultivation site, and farmer's ability to match the art and science of algal farming. Nori cultivation has certain advantages over other forms of mariculture as it does not require large amounts of capital investment. Unlike other aquaculture crops, nori reaches a harvestable stage in several weeks as compared to years for fin and shellfish industries.

Fishing village receptivity to adopting algal farming as a new commercial pursuit is enhanced by the presence of, or communication by, successful algal-farming villages or entrepreneurs; availability of experienced extension agents with nori farming training courses; and the support of commercial or academic community. Fishermen can make good aquaculturists but they need training and a mindset shift. Commercial capture is philosophically different from husbandry. Seaweed farms take daily care and attention to detail. Team work and cooperation are tantamount to success.

December 1, 2009

Ira A. Levine
Dinabandhu Sahoo

Acknowledgements

This work in part was supported by the U.S. State Department's Fulbright New Century Scholar Program, Maine-New Hampshire Sea Grant College Program, National Coastal Research Institute, University Southern Maine, Lewiston Auburn College and the University of Delhi, Graduate Botany Department. Additional material has been extracted from our translation of the Zen Gyo Ren, Zen Nori Coop farming publication. IAL thanks are extended to Dr. Hitoshi Kito, formerly of Shimonoseki State Fishery University, Dr. X. G. Fei, formerly of Academica Sinica, Mr. Urano of the Seto Inland Sea, Dr. Thomas Mumford, Andrew Stevenson, and Steven Crawford for all their efforts to transfer technology from Japan, China, Washington State to Maine and assistance on the Nori Farming and Teaching Nori Cultivation Extension Course manuals. Several of the artist's renditions of knots, anchors, & system configurations were adapted from Byce et al., 1984. Equipment for nori farming in Washington State Volumes 1 and 2. I would like to extend a belated thanks to Drs. Maxwell Doty and Harold Humm, two giants who are surely missed. Lastly, I would like to thank my wife Laurie Lee W. Levine for her tireless support of both my career and the development of this book. Her indexing of files and photos, along with her long-distance support of the manuscript preparation has been invaluable.

D. Sahoo thanks Professor Masao Ohno from Usa Marine Biological Institute, Kochi University, Japan for providing invaluable assistance during his stay in Japan as an INSA-JSPS Fellow. DBS also thanks Professor Charles Yarish from University of Connecticut, Stamford USA for extending unconditional support during his stay in his laboratory under a visiting Fellowship and use of several photographs. In fact, I learnt a great deal about *Porphyra* during my Stamford days. I thank my wife, Dr. Meenakshi Munshi for her support. DBS thanks his former student Dr. Sunita Singh who completed her Ph.D. in *Porphyra* and for providing valuable information in the preparation of this book.

We have received much help from Dr. Pooja Baweja, M. Arunjit Singh, Elangbam Geetanjali, Savindra Kumar, Salam Sonia Devi, Gaurav Kumar, Nitin Pipralia, Arvind K. Bhardwaj, Priyanka Verma and Vivek Chopra research students of Marine Biotechnology Laboratory, Department of Botany, University of Delhi, Delhi-110007 India during the compilation of this book.

We also thank our publisher, I.K. International, for agreeing to this project and publishing our book in such a timely manner.

Ira A. Levine
Dinabandhu Sahoo

Contents

전남김
全南産の海苔
JeonNam Laver
무공해 자연식품 전남
맛과 영양도 최고입니

1 Introduction

Figure 1. A Korean *Porphyra* advertisement. (Courtesy: D. Sahoo, India)

INTRODUCTION

Porphyra, a red seaweed or red marine algae is one of the most commercially important crops. This sea plant or sea vegetable can be called a wonder plant and one of the best gifts of nature. The plant is eaten either in dried or processed form all over the world. Today this plant is a multibillion dollar industry and can be called "Gold from the Sea" whose annual food value approaches US$ 2 billion. In fact one sheet of *Porphyra* weighing approximately 3 grams can cost up to US$ 10 and the handmade "Emperor's Nori" can reach US$ 15 per sheet. The 2008 harvest data reveals approximately 21.6 billion nori sheets were processed and the demand annually increases as the awareness of its flavor and nutritional value grows.

Porphyra, commonly known as: **Laver** – UK, USA, Canada; **Karengo** – New Zealand; **Kim** – Korea; **Nori** – Japan (most commonly referred to); **Purple Laver** – Britain and Ireland; and **Zicai** – China (Figure 1), has been recognized as an important source for food and medicine since its earliest recorded use by Chi Han in China in 300 B.C. Today, the Japanese, Chinese, Koreans and other people of Asia and the western Pacific continue to value nori as an integral part of their diet (Figure 2).

Nori contains high levels of protein (25-50%), vitamins (greater percentage of Vitamin C than in oranges), trace minerals, and dietary fiber (Table 1) (Noda, 1993; Sahoo, 2000; McDermid

Figure 2. Roasted *Porphyra* sheet with the United Kingdom and United States' traditional name of "Laver" used on the packaging. (Courtesy: D. Sahoo, India)

and Stuerck, 2003). The plant contains 17 types of free amino acids including taurine, which controls blood cholesterol levels (Tsujii *et al.*, 1983). Nori tea is actively relied on in the treatment of stomach cancers and ulcers whereby the Chinese refer to it as the happy gift from the sea.

Table 1. *Porphyra:* nutritional constituents per 100 grams fresh weight.

Nutrients	Amount present in *P. abbottae*
Food Energy (K (Cal)	279
Protein (g)	24.2
Fat (g)	1.4
Carbohydrate (g)	58.0
Crude fiber (g)	25.2
Ash (g)	16.1
Thiamine (mg)	0.37
Riboflavin (mg)	1.79
Niacin (mg)	6.7
Vitamin C (mg)	11.6
Vitamin A (Retinol equivalents)	263
Calcium (mg)	230
Phosporus (mg)	474
Sodium (mg)	3300
Potassium (mg)	3140
Magnesium (mg)	623
Copper (mg)	1.7
Zinc (mg)	1.7
Iron (mg)	10.5
Manganese (mg)	1.6

(Courtesy: D. Sahoo, India)

2 Value and Quality

Porphyra is one of the most valued maricultured seaweeds in the world (Arasaki, 1982; Miura, 1975; Okazaki, 1971; Oohusa, 1993a, b; Tseng and Fei, 1987) with total global sales exceeding U.S. $ 1.5 billion. In 2008, approximately 21.6 billion sheets (~ 65,000 dry metric tons or 1,810,000 wet metric tons) of *Porphyra* were maricultured.[1] The three major nori farming countries; Japan, Korea, and China produced 8,980,000,000, 8,600,000,000, and 4,000,000,000 sheets of nori in 2008, respectively. This number principally focuses on nori as a food product. The nori market growth potential is significant due to its utilization diversity but its use as a traditional holiday gift staple has declined due to changing economic demographics. Absolute global nori harvest and utilization data is challenging to find due to its use as a component ingredient, e.g. recipes, vitamins, neutraceuticals, cosmeceuticals, hair products, and dyes.

The Japanese originally imported nori from

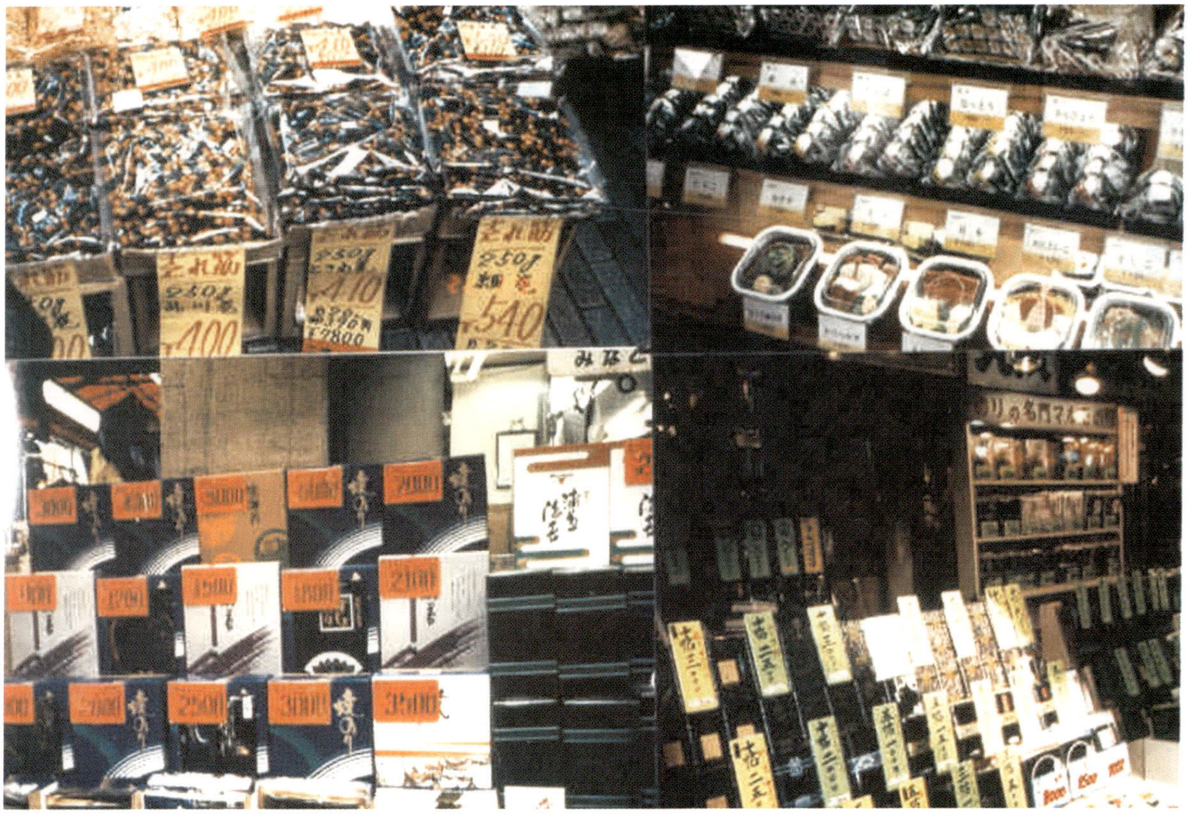

Figure 3a. Ueno, Japan Nori shop with full range of quality products ($ 1.00-70.00 per package). (Courtesy: I.A. Levine, USA)

China and have become nori connoisseurs and dominate its global cultivation and marketing. The Japanese people have integrated many different applications for the sea vegetable into their daily lives and have established a rigorous set of standards by which to evaluate nori grade and quality. Nori, like cigars, can range in price from a minimum of $ 0.03 per sheet (anything less is destroyed to prevent price devaluation) to over $ 10.00 per sheet (Figure 3a).

The price at auction and subsequently at the retail level is established through a highly detailed rating system. Japanese nori is rated from grades 1-155 by one of the country's ~ 200 official nori grading officials and the farmer's sheet ratings are the basis for biding at the auction. Korean and Chinese nori are not subjected to equivalent grading rigors. Currently, they are graded from A-F by the producer who typically distributes the harvest to nori marketing companies. The grade and quality of the nori ultimately determine its final use. The highest quality nori is set aside for consumption while the lowest qualities are alternatively used as pet foods of cosmeceutical ingrediently.

Figure 3b. Nori products available in Korean market. (Courtesy: D. Sahoo, India)

3

Uses of *Porphyra*

FOOD

Dried nori sheets are packaged and are available throughout retail distribution markets and normally purchased by homemakers or cooks (Figure 4) and hand toasted over a stove or hotplate to guarantee freshness of the crispy sheet. The toasted nori sheet is offered with rice at each meal throughout the day.

Toasted or roasted nori sheets are commercially prepared at the nori marketing company's processing and packaging factories. The purchased dried nori sheets, stored in temperature controlled warehouses, are unpacked, redried and prepared for roasting. The sheets are rapidly run through a roasting conveyor system, cooled and immediately sealed in the retail packaging. The newly roasted nori is packaged

Figure 4. Dried nori sheets available in Japanese market. (Courtesy: D. Sahoo, India)

into cases and shipped to several distinct market distribution centers (Figure 5a). The roasting factories have developed flavoring processes and numerous sheet sizing configurations aimed at the fast food and younger generation markets (Figure 5b). Alternatively, Chinese *Porphyra* preparation consists of a flattened cake of nori blades made without the use of a sheet making machine, which is used as a flavor ingredient in soups.

The infusion of Chinese nori to the world market in the last 20 years has resulted in a significant drop in the retail sheet price. Ten sheets of Chinese nori in an oriental grocery store range from $ 1.60 to 2.00. Korean nori is priced in the $ 2.50 - 3.50 per ten-sheet package. Japanese nori, although difficult to locate in retail markets, is up to $ 1.00 per sheet and is usually distributed to the higher end sushi bars and Japanese restaurants in 50 to 100 sheets packages. A private US-based marketing report identified four primary retail distribution outlets for nori; natural food stores, gourmet/specialty food stores, supermarkets, and Asian/oriental groceries with > 25 brands of nori sheets penetrating these market segments.

The world nori market is currently dominated by Zen Nori (Socio-economics society of nori culture development) along with Zengyoren (National Federation of fisheries Co-operative Associations) the national nori farming and marketing cooperatives and the large nori marketing companies: Koasa Trading Co., Ltd., Yamamoto Noriten Co., Ltd, Yamamotoyama Co., Ltd, Ohmoriya Co., Ltd., Yamagataya Nori Co., Ltd, Takaokaya Co., Ltd. These companies represent the overwhelming majority of global nori sheet purchases.

PIGMENTS

Porphyra has been the principal commercial source of the fluorescent pigment r-phycoerythrin (Figure 6). R-phycoerythrin is utilized as a fluorescent "tag" for immunofluorescent

Figure 5a. Typical nori counter filled with different delicious delicacies. (Courtesy: D. Sahoo, India)

Figure 5b. Newly developed rice and nori triangular snack packaging. (Courtesy: D. Sahoo, India)

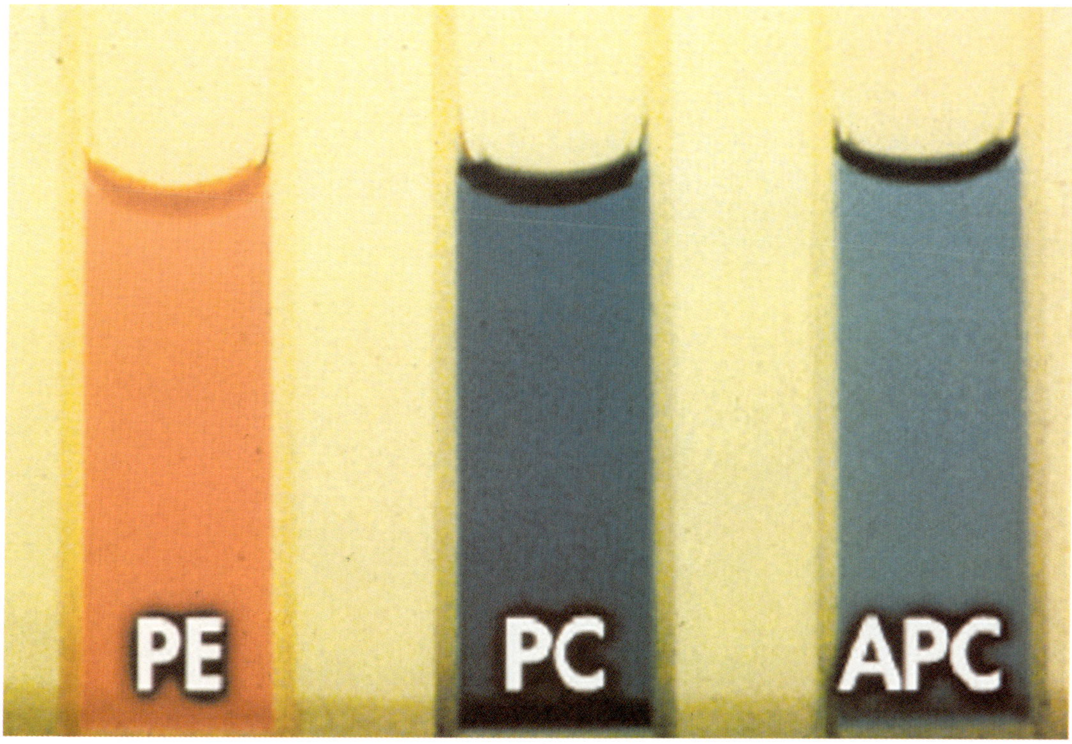

Figure 6. *Porphyra* dyes: Phycoerythrin PE, Phycocyanin PC, & Allophycocyanin APC. PE used in fluorescent labeling. (Courtesy: I.A. Levine, USA)

diagnostics, flow cytometry, and antibody labeling (Mumford and Miura, 1988). R-phycoerythrin is globally produced and marketed by over 40 companies and has a market value greater than $12 million per year. The approximate value of *Porphyra yezoensis* as a raw material source of r-phycoerythrin is $40-45/kg fresh weight (fob). *P. yezoensis'* r-phycoerythrin C-1 experiences optimal absorbance and fluorescence maxima desired by the industry.

BIOREMEDIATION

Porphyra is utilized as a nutrient scrubber and water purifier due to its absorbance of excessive nutrients in water polluted with human or animal effluents while producing algal biomass and secondary products.

The production of finfish, shrimp and prawns in pond or cage based monoculture systems results in the production of effluents high in organic and inorganic nutrient concentrations. As the costs of environmental impacts are passed along to the finfish cultivators, the ability to raise a second revenue-producing crop, i.e. nori, while mitigating nutrient enrichment presents an attractive alternative to aquaculture companies (Figures 7 a & b).

Figure 7. A) Multitrophic aquaculture; Cobscook Bay, Maine, USA. Salmon cages left and right with nori nets just off to the side of the salmon cage. (Courtesy: I.A. Levine, USA)

Figure 7. B) Multitrophic aquaculture nori net. Deep, dark color due to fish effluent laden waters. (Courtesy: I.A. Levine, USA)

Multitrophic aquaculture systems generate biomass and secondary products while producing measurable improvements in:

biological oxygen demand	dissolved oxygen
phosphate	nitrate & ammonium
trace metals	sludge accumulation
total organic matter	agrochemicals
organic pollutants	pathogens

Farmed fish are fed nutrient-rich feed and inevitably, a sizable portion is not consumed and contributes to waste, along with excrement. The nitrogen loading from a salmon farm of 200,000 fish, for example, has been estimated to be roughly equivalent to the untreated sewage of 20,000 people (Hardy, 2000). Nori production nets were assembled adjacent to salmon pens to determine its ability to absorb salmon effluents. Measurements of tissue nitrogen and phosphorus were made in three *Porphyra* species (Figure 8) in northeastern Maine, USA and compared to ambient water nitrogen and phosphorus levels. All three species (*P. purpurea, P. umbilicalis* and *P. yezoensis*) showed an ability to concentrate nitrogen and phosphorus with the commercial species *P. yezoensis* exhibiting the highest levels of effluent accumulation. *P. yezoensis* accumulated nitrogen up to 72 mg/g DW compared to ambient concentrations of 11 mM or less and phosphorus levels around 8 mg/g DW compared to ambient concentration below 1 mM, (Chopin et al 1999).

Benefits of using *Porphyra* to bioremediate hatchery and finfish farms include:

- Fast growth rate over the range of desired temperatures
- Ideal morphology for nutrient uptake; uni- and bi-serate blades with high surface area to volume ratio
- High rates of N and P uptake
- Established system for producing blades which can be grown in tank cultivation,

Figure 8. Aerial photograph of multitrophic aquaculture: *Porphyra*: salmon pilot farm, Maine USA. (Courtesy: I.A. Levine, USA)

- Easy harvesting,
- The ability to grow at a wide range of salinities
- The ability to produce a commercially valuable form of r-phycoerythrin

OTHER USES

In Europe, nori has been used in herbal medicines or nutraceuticals since the time of the Greek and Roman Empires. Nutraceuticals are a recently described class of foods and food ingredients that serve to provide specific health benefits for the prevention or treatment of particular diseases (Figure 9). *Porphyra* and other algae have been used in Asian countries for years as a source of trace minerals and general health enhancer. This contrast with regular foods that are consumed largely for their nutritional value (e.g. essential amino acids, vitamins, minerals) and with traditional pharmaceutical products which are administered in pill or injection form and focus largely on healing and treatment. Nutraceutical ingredients in their simplest forms are often micronutrients present in normal fruits and vegetables which need to be taken in higher doses or in more readily bioavailable forms to exert their beneficial effects.

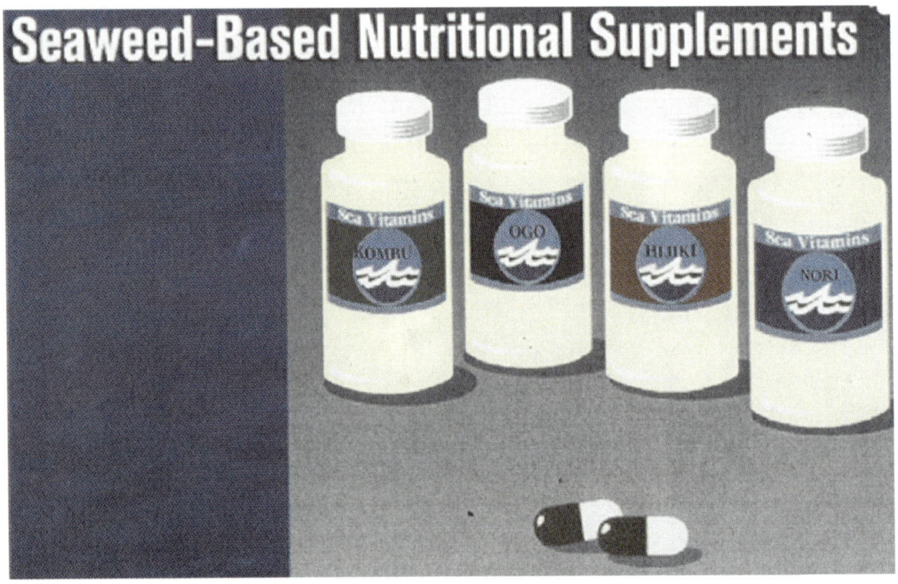

Figure 9. Seaweed-based nutraceuticals. (Courtesy: I.A. Levine, USA)

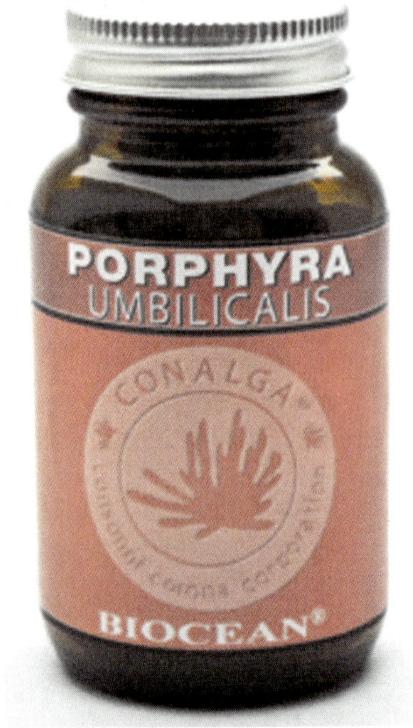

Porphyra based nutraceuticals. (Courtesy: www.algheria.it/schede/acidi-grassi-polinsatu...)

4 Biology

The genus, *Porphyra* was first established by C.A. Agardh in 1824. *Porphyra* has a heteromorphic life history, which alternates between a gametophytic phase (blade) (Figure 10 a) and a filamentous sporophyte called conchocelis phase (Figure 10b) (Drew, 1949; Miura, 1975; West and Hommersand, 1981; Sahoo *et al.*, 2002 & 2006). This diploid conchocelis phase

Figure 10 a. Gametophytic blade of *Porphyra vietnamensis.* (Courtesy: D. Sahoo, India)

Figure 10b. Sporophyte of *Porphyra vietnamensis* conchocelis phase.

in the life history was earlier thought to be *Conchocelis rosea*, a shell-boring organism. However, Drew (1949) demonstrated in *P. umbilicalis* (L.) Kutzing that conchocelis phase is the sporophytic phase in the life history of *Porphyra* and not an independent shell-boring organism.

DISTRIBUTION

Porphyra species normally grow on rocks (Figure 11) or as epiphytes in the intertidal or shallow subtidal zones and are generally seasonal in their appearance and growth.

Porphyra species are widespread in temperate water regions and are particularly abundant in North Pacific Ocean (Mumford and Cole, 1977). More than 133 species of *Porphyra* have been reported from all over the world (Figure 12), including 28 Japanese species, 30 species from North Atlantic coasts of Europe and America and 27 species from the Pacific coast of Canada and United States (Yoshida et al., 1997). Seven species have been reported from tropical waters of Indian coast (Sahoo *et al.*, 2001). However, only six species, i.e. Kjellman, *P. haitanensis* Chang and Zheng, *P. pseudolinearis* Ueda, *P. kunidea* Kurogi and *P. seriata* Kjellman are being commercially cultivated in China, Japan, and Korea (Sahoo, 2005; Sahoo and Yarish, 2005). The first two species, *P. yezoensis* and *P. tenera* comprise nearly 90% of the global harvest (Jensen, 1993; Oohusa, 1993; Tseng and Fei, 1987).

Figure 11. *Porphyra* growing on the rocky shores. (Courtesy: D. Sahoo, India)

MORPHOLOGY

The genus *Porphyra* has a simple morphology; the plants are either round, ovate, obovate, linear or linear lanceolate. Different species vary greatly in their texture, color, and size (Figure 13). Blades are usually thin and smooth with undulating margins. Some *Porphyra* species have microscopic denticulation on the margins of the blade. Thalli thicknesses are highly variable: generally, the fertile sector of the thallus is much thicker than the vegetative portion and changes according to their age with younger plants generally thinner. Individually, the sterile margin is thinner as compared to the thicker basal portion. Four species groupings are recognized by thalli thickness: 1) thin, less than 40 µm, 2) medial, 40-60 µm, 3) thick, 60-80 (-100) µm, and 4) very thick, more than 100 µm (Yoshida, 1997).

Individuals can also have blades that may be divided into male and female sections (Tang and Fei, 1999) or have sectored morphology (Mitman and Van de Meer, 1994). The plants can grow from 5 to 35 cm in length. The annual blades are membranous, monostromatic or distromatic, monoecious or dioecious. Alternatively, the shell boring life history conchocelis phase is perennial (Hafting, 1998).

Figure 12. Global distribution of *Porphyra.* (Courtesy: D. Sahoo, India)

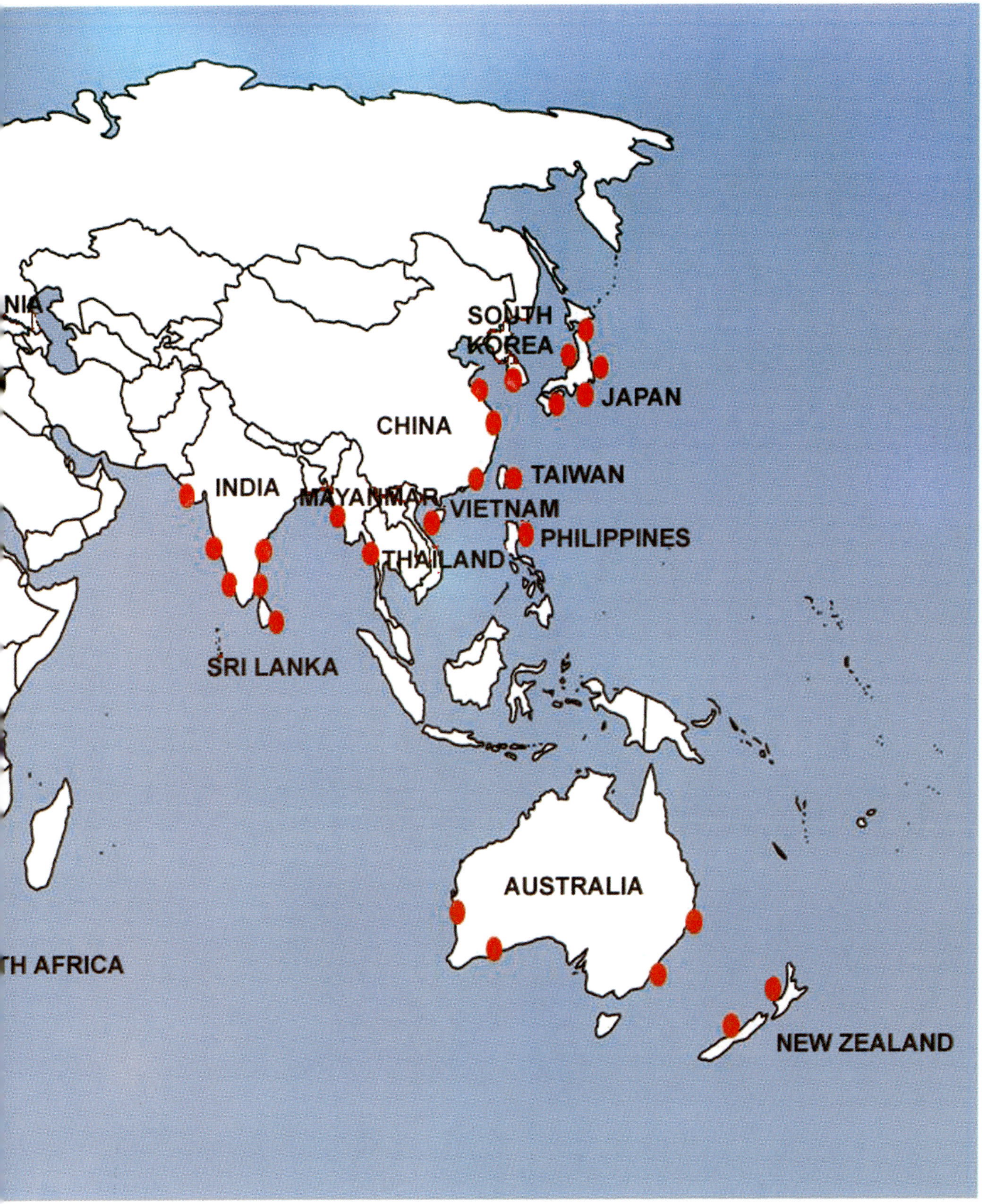
SOUTH KOREA
JAPAN
CHINA
TAIWAN
INDIA
MAYANMAR
VIETNAM
PHILIPPINES
THAILAND
SRI LANKA
AUSTRALIA
TH AFRICA
NEW ZEALAND

Figure 13. Variability of *Porphyra* blade forms. (Courtesy: D. Sahoo, India)

CYTOLOGY

Number of cell layers and chloroplast per cell are important characters for identification of *Porphyra* species (Figure 14a,b,c). Based on these characters there are three types of structures which are used to distinguish various species (i) monostromatic thallus with one chloroplast per cell (ii) monostromatic thallus with two chloroplasts per cell or partly distromatic (iii) distromatic thallus with one chloroplast per cell. Most of the species of *Porphyra* are monostromatic, membranous and possess one chloroplast per cell such as *P. argentinensis* sp. nov., *P. pujalsii* (Piriz, 1981), *P. suborbiculata, P. linearis, P. pseudolinearis, P. purpurea, P. fucicola, P. leucosticta, P kurogii* (Lindstrom and Cole, 1992), *P. adamsiae, P. cameronii, P. subtumens, P. woolhousiae, P. kaspar* (Nelson, 1993) etc. Only few species are distromatic such as *P. variegata, P. miniata, P. leucosticta, P. amplissima, P. cuneiformis* (Lindstrom and Cole, 1992).

Antennae pigment manipulation has yielded color mutants within one species, *P. yezoensis.* Color variants can alter growth rates and develop unique finished nori sheets (Figure 14d and e).

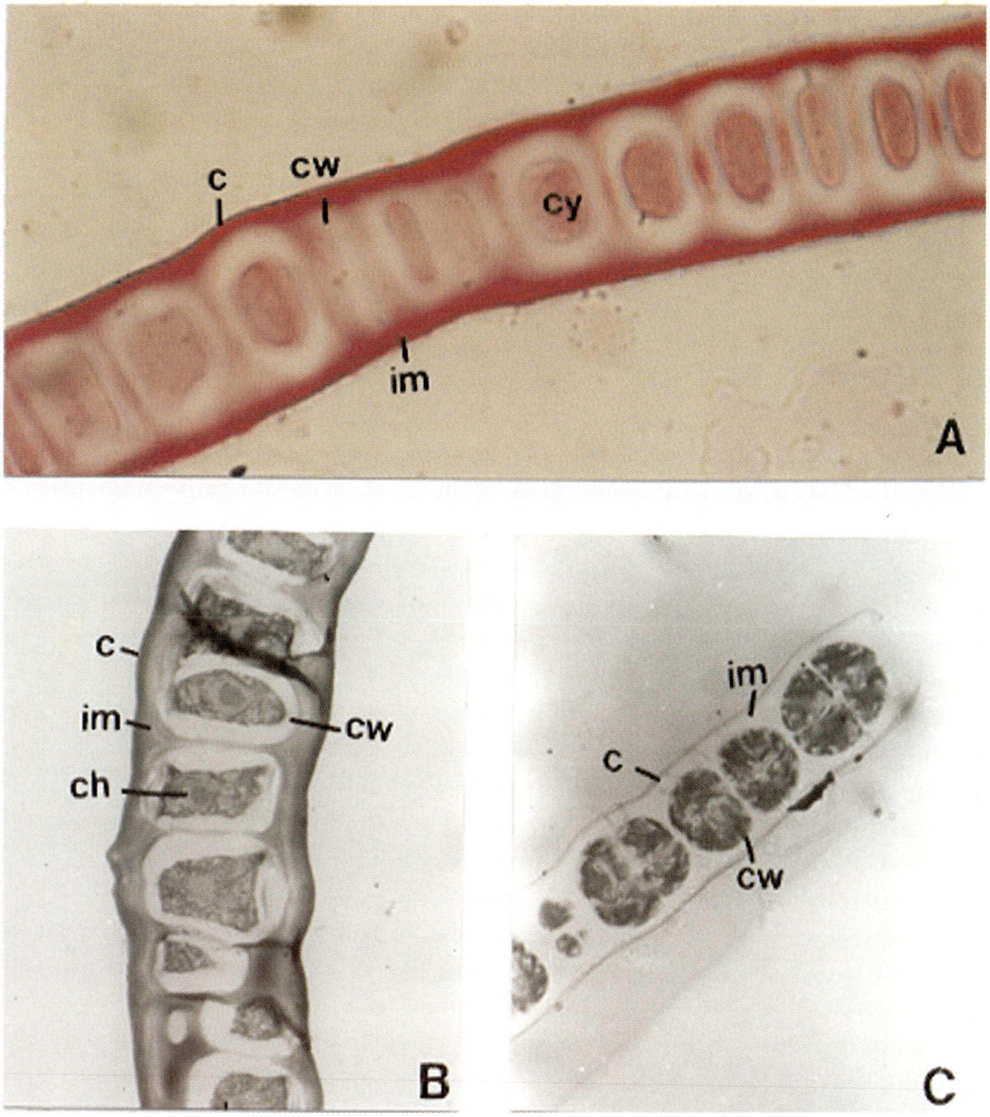

Figure 14. Photomicrograph of cross section of *Porphyra* uniserate layer. (c – cuticle, ch – chloroplast, cw – cell wall, im – intercellular matrix, cy – cytoplasm. (Courtesy: Sunita Singh, India)

Figure 14. (d) Emerging haploid color mutants blades (Courtesy: Y. Xinghong)

Figure 14. (e) Fully formed haploid color mutants (Courtesy: Y. Xinghong)

i. Reproduction

In monoecious species, patches of male and female cells are mixed at the blade margins (e.g. *P. tenuipedalis* and *P. lacerata*) or the thallus is chimeral into male and female portions (e.g. *P. variegata*) (Figure 15).

In dioecious species, male and female cells are produced on different foliose thalli (e.g. *P. dentata*) (Notoya, 1997). The males are pale yellow and the females dark-red in color. The blade's color range from rose-pink to red, yellow, green and brown when found in the intertidal area.

Figure 15. Monoecious chimeral *Porphyra.* (Courtesy: I.A. Levine, USA)

Asexual reproduction by archeospores are formed singly from vegetative cells or by repeated partition of vegetative cells and released from the margins (Figure 16 a & b).

In sexual reproduction, certain mature vegetative cells become differentiated into female gametangia and others on the same or different thallus become differentiated into colorless male gametangia. After fertilization, the female gametangia divide to form packets of spores called zygotospores (Figure 17 a & b). After release, the zygotospores usually germinate unipolarly to produce the filamentous conchocelis phase. The conchocelis can survive in adverse environmental conditions, but give rise to conchosporangia and conchospores under

Figure 16. A) Release of archeospores from the mature thallus.

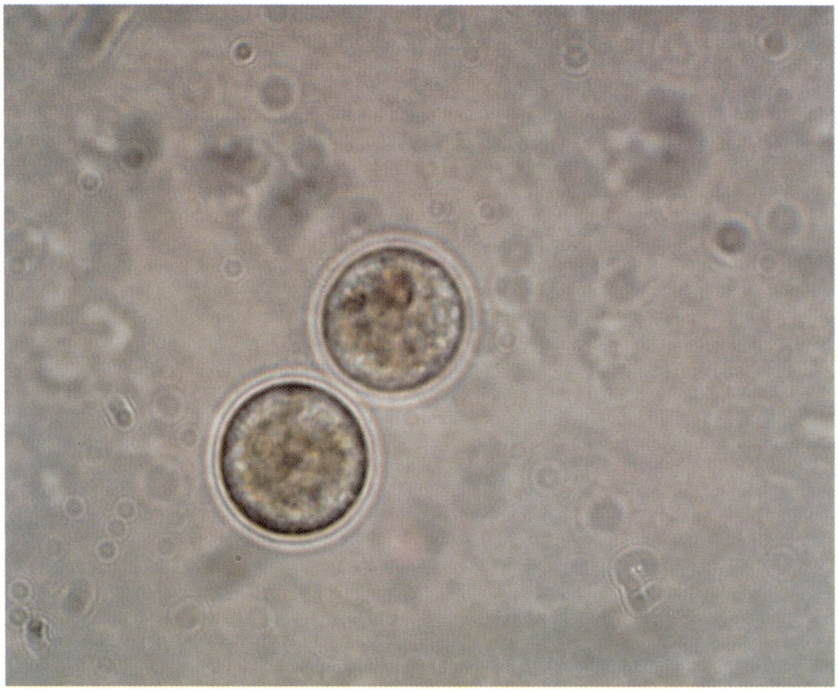

Figure 16. B) Released archeospores.

suitable conditions. The conchospores germinate by bipolar modes to give rise to young chimeric thalli (Sahoo *et al.*, 2002).

Porphyra has a biphasic life history which commences from spore to blade in a seasonally controlled manner. Completing the life history revolutionized the *Porphyra* farming industry as both phases were able to be maricultured (Figure 18).

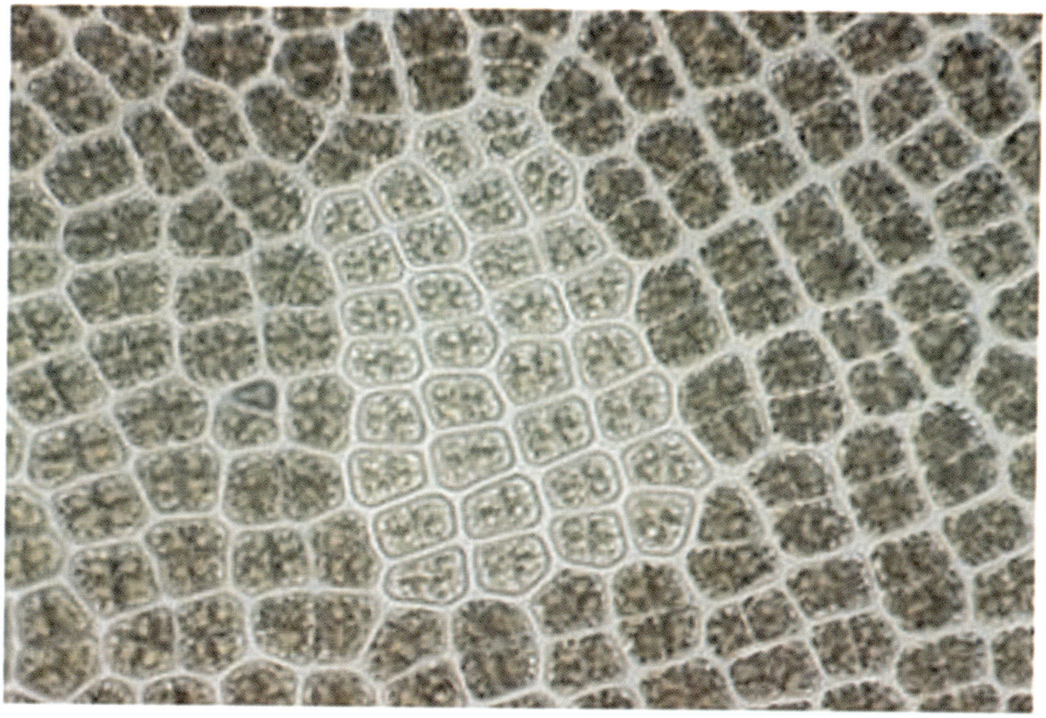

Figure 17. A) *Porphyra* spermatangia and zygotosporangia.

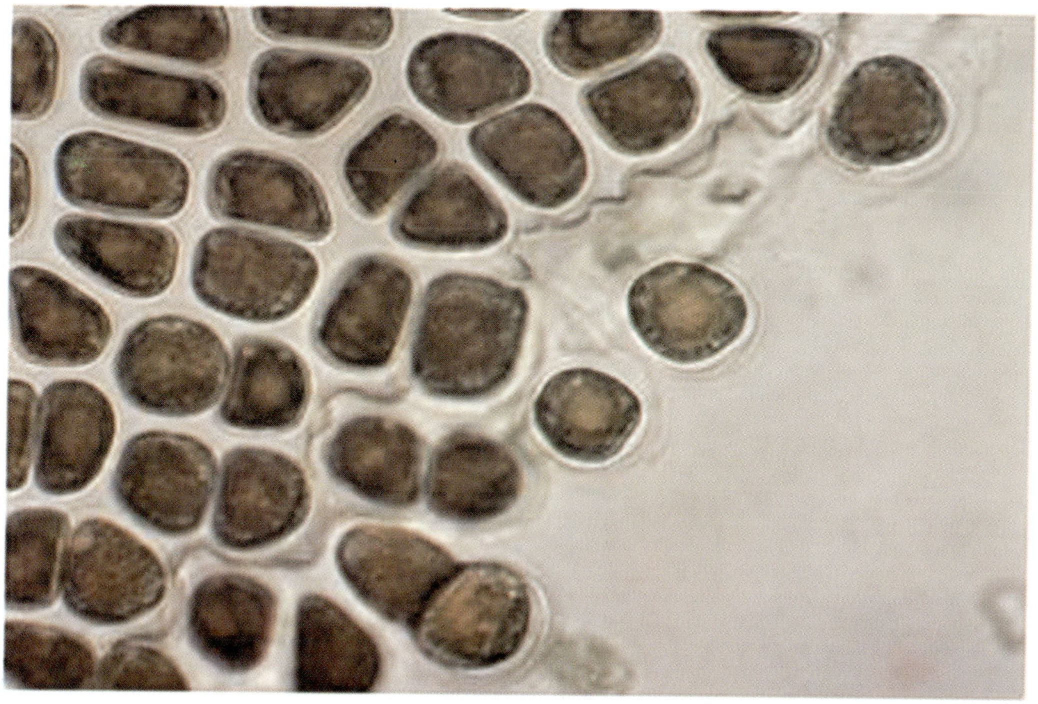

Figure 17. B) Release of zygotospores

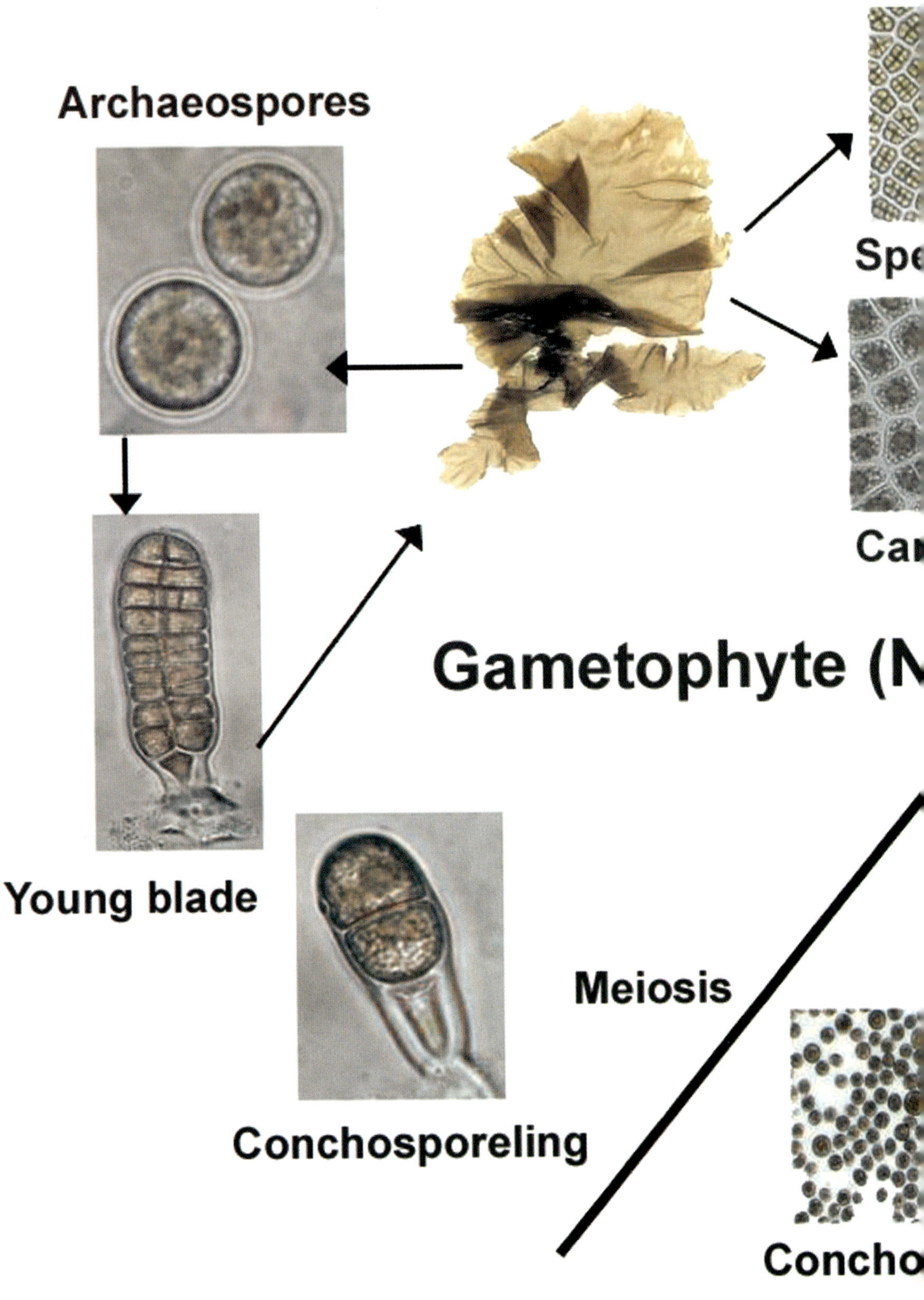

Figure 18. *Porphyra* life history illustrating different stages of development (reproduced from Sahoo *et al.*, 2002).

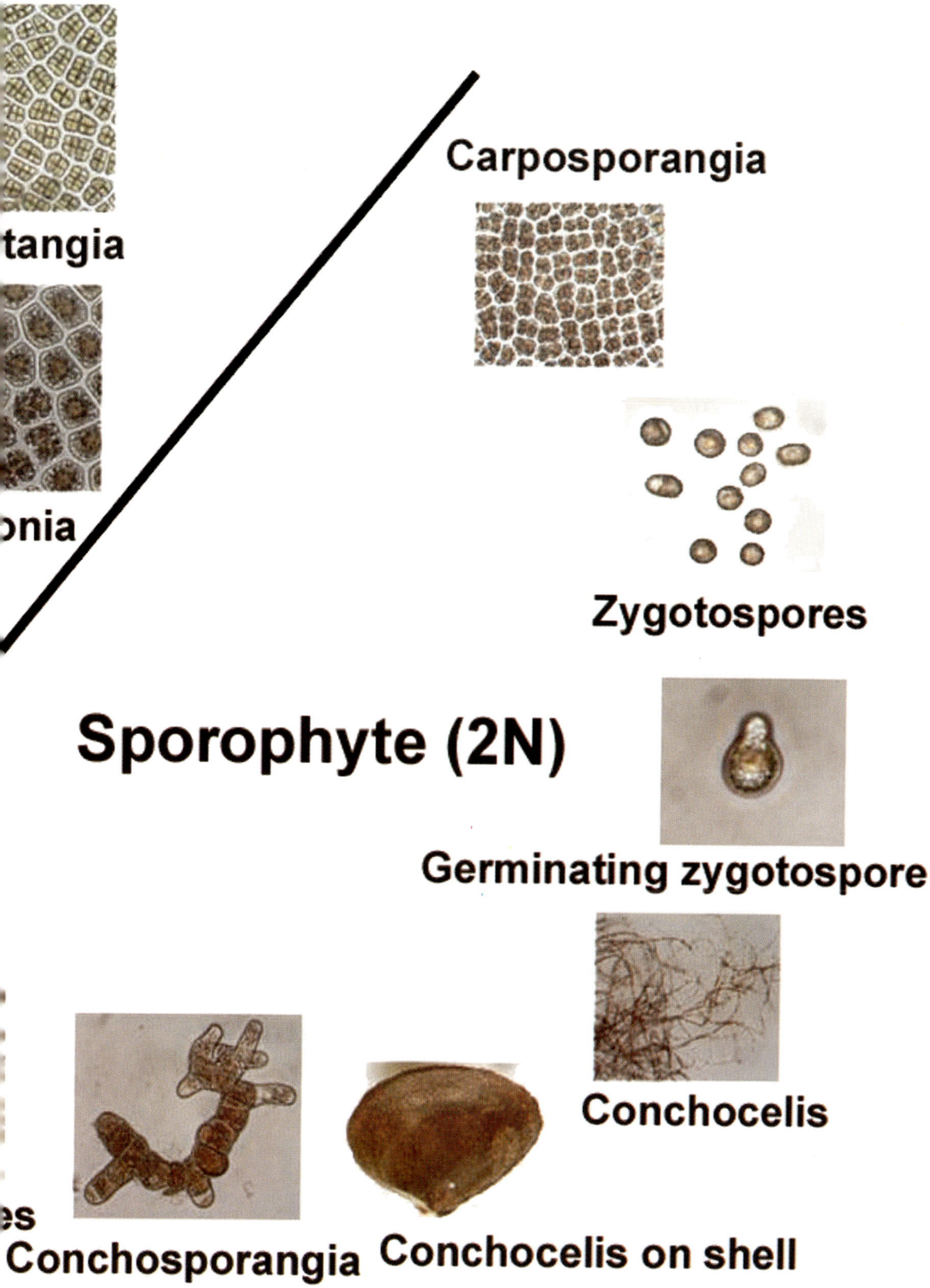
Carposporangia
Zygotospores
Sporophyte (2N)
Germinating zygotospore
Conchocelis
Conchosporangia
Conchocelis on shell

5 Farming

FARMING HISTORY

Japanese *Porphyra* cultivation began in Tokyo Bay, near the estuary of the Sumida River, in the 1650's with the placement of bamboo twigs and branches at the correct tidal height. Plants grew on the branches in the fall and early winter, and were hand-picked (Figure 19). This method of seeding twigs was common until the twentieth century when methods that made use of nets became popular (Mumford, 1988).

Figure 19. Classic print of ancient nori farming practices.

Modern techniques for *Porphyra* cultivation were introduced in Japan – 1960's, Korea – 1980's and China – 1990's involving five major steps : (1) culture of conchocelis, (2) seeding of culture nets with conchospores, (3) nursery rearing of sporelings, (4) production system cultivation and (5) harvesting.

The nori industry in Japan is not vertically integrated. The traditional nori farming model includes companies that; culture the conchocelis which will seed nets or sell shells to farmers; family farmer (husband, wife and children) who will farm 200 nets per year (Figure 20); village processing cooperative where the nori is dried into sheets; the family farmer or cooperative's sheets will go to auction. The auctioned nori is then roasted and packaged by the nori trading companies which then enter into the distribution and retail markets.

The number of Japanese nori farmers have declined from approximately 27,000 in the 1970's to 4,868 in 2008 with a 16.5 % decrease in just the last year; due in part to socio-economic changes in Japanese society and the decline in the demand for high quality nori sheets. The 82% decrease in Japanese nori farming fishermen have resulted in only a 18% drop in the number of nori sheets produced due to the advent of *Porphyra* farming mechanization. The traditional nori family farmer is giving way to fishery or village cooperatives (Figure 21 a & b). The development of corporate nori farming operations and processing concerns, with salaried employees, seed, culture, harvest and process all their own

Figure 20. Typical individual Japanese nori farming family with 200 nets to fixed pole farm; Ariake Bay, Japan. Nori nets laden with biomass, husband and wife harvest the crop and the baskets loaded on family nori fishing boat.

Figure 21. Chinese nori fishing cooperative. A) Village cooperative traveling out to the nori culture grounds on extended intertidal mudflat. B) Unique motorized cart transport workers, equipment and supplies to mud flat culture grounds.

nori. Finally, there are attempts to vertically integrate the nori process chain to include the farming and processing of the sheets, but also to roast, package, distribute and sell directly into the retail value chain.

Additional changes to the industry have taken place in Korea and China over the past 25 years. Japanese nori trading companies have been very aggressive in developing Japanese nori sheet style production in Korea and China. The Chinese crop is variable with quality not yet approaching that of traditional Japanese production systems but the cultivation economics of China are an order of magnitude less than Japanese nori production costs.

American Experience in Nori Cultivation

Commercial cultivation of the red alga *Porphyra yezoensis* was initiated in Maine, USA in 1991 by Coastal Plantations International, Inc. (CPI) (Levine and Crawford, 1997; Levine, 1997a, b) following Washington State's efforts of the 1980's. The transfer and modification of Japanese, Korean, Chinese and Washington State's nori cultivation and processing technologies to Maine resulted in the world's first organically cultivated and processed nori (Figure 22) in addition to the establishment of a nori:finfish multitrophic aquaculture system. The nori farming effort reached a maximum of 1000 saku (Japanese nori farming term to indicate a net space in a production system array) with two nets/saku/season. Production values were inconsistent, reaching a maximum of 8000 sheets per net. The farm's processed nori was distributed to health food stores, direct mail order to Japan (Figure 23) and as a source of fluorescent labelers marketed directly to biomedical concerns.

Figure 22. Ira Levine farming nori in Cobscook Bay, Maine USA

Figure 23. Maine Nori Company's dried nori sheets.

Canadian Experience in Nori Cultivation

Following the introduction of nori farming to Washington State in the mid-1980's, approvals were granted for the introduction of *Porphyra yezoensis* and frozen nori nets to Victoria, British Columbia in late 1980's. Nori was farmed approximately three years before the effort collapsed due to lack of reaching critical production values. In 1991, nori farming started in the Canadian Province of New Brunswick in Harbor De Lute adjacent to an existing salmon farming operation. The introduction and assembly of a small pilot test farm was initiated and growth rates realized indicated a real potential for commercial operations in the Canadian Maritime Provinces.

Indian Initiative in Nori Cultivation

In recent years, efforts have been made to cultivate *Porphyra* in India. The co-author initiated a major program on "Marine Algae Cultivation

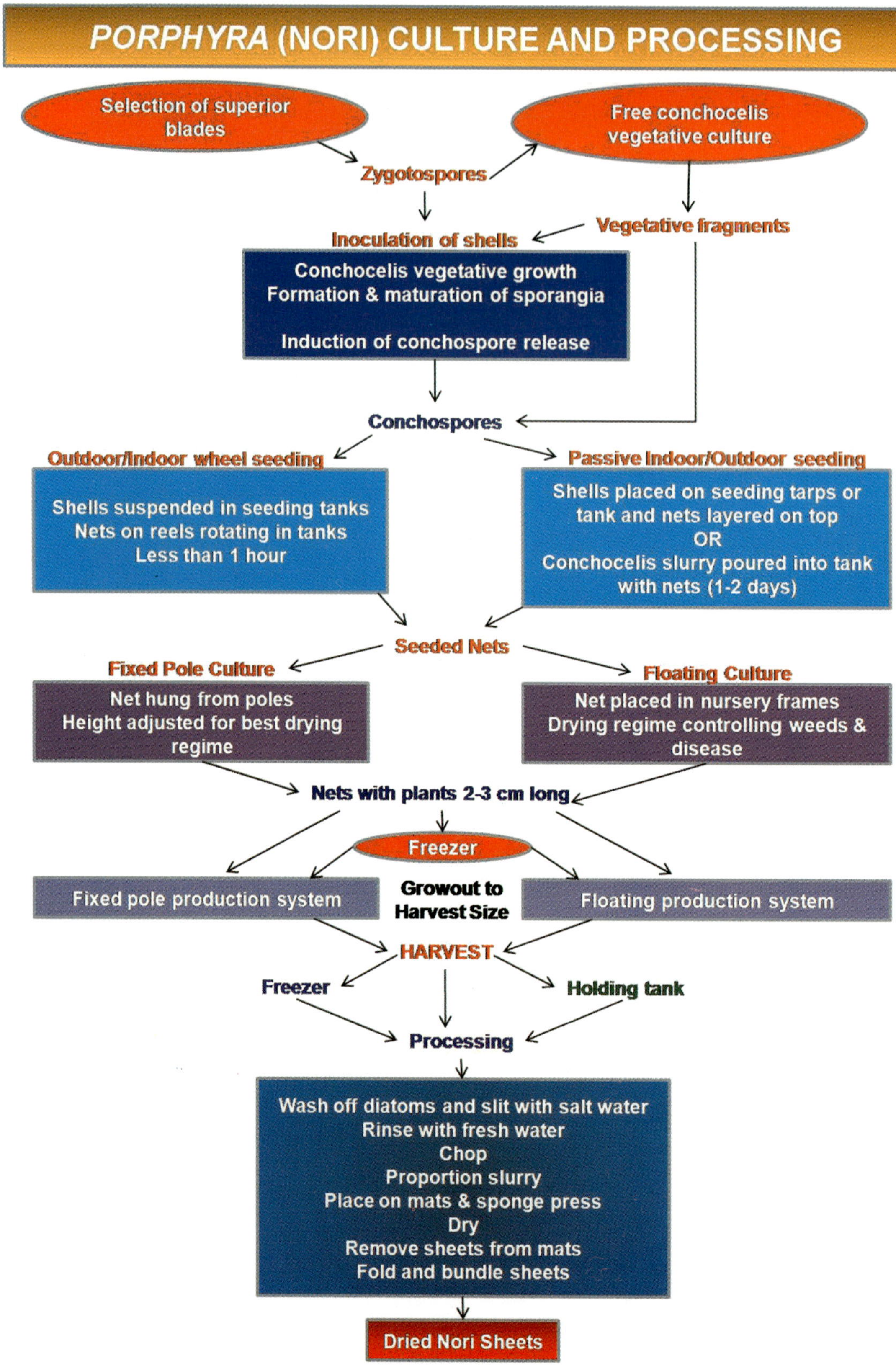

Figure 24. Entire nori farming process stream, from cultivar selection to nori sheets. (Courtesy: D. Sahoo, India)

and Processing for Livelihood Generation in Coastal States and Union Territories of India" with the assistance of the Department of Science and Technology, Government of India, New Delhi.

Figure 24 illustrates the complete nori farming enterprise. The remainder of this chapter details the steps outlined in this flow diagram.

TYPES OF FARMING

There are two major forms of *Porphyra* (nori) farming which are discussed in greater detail: Open water nori farming which takes on two forms; fixed and floating (a hybrid semi-floating system has also been developed and utilized in massive Chinese mud flat nori cultivation grounds) (Figure 25 a,b,c) and Land-based nori farming.

NOTE:

The remainder of the book is organized as a series of stepwise training modules for the benefit of aquafarmers and aquaculture extension agents. The unique format is intended to be used in the classroom, laboratory, conchocelis land-based farms, seeding facilities, nursery and production systems, harvesting platform, and land-based processing facilities.

OPEN WATER NORI FARMING

Tailoring:

OBJECTIVES: 1) To progress from the known to the unknown, as in; to understand "Farming" and then to begin to understand "Nori Farming".

Figure 25a. Nori farming open water systems.

(b)

(c)

Figure 25. Nori farming systems: B) Open water semi-floating C) Fixed pole.

2) To create an awareness as to the personal traits needed to succeed at farming and therefore nori farming. 3) To demonstrate the diversity that currently exists in farming and aquaculture.

RATIONALE: People have expectations, likes and dislikes. There is a work ethic required of people who choose to farm. There are characteristics of "farmers". The time to eliminate misconceptions and clarify those traits and characteristics that will lead to success in nori farming is in the beginning of the farming effort. By documenting the diversity that exists today in farming/aquaculture you further validate seaweed/nori farming as a real type of farming.

The nori requires constant tending and cultivation success is dependent on both the discipline and nature of the farmer along with the vagaries of the weather as with most forms of agriculture.

READING MATERIALS AND ACTIVITIES:

1) Define farming as a group and put the group definition on the board.
2) Define a farmer and put the group definition on the board. Follow up question like; what traits or characteristics make a farmer good at what they do?
3) The extension agent should prepare a folder for each farmer with articles about farming in general. The articles should cover a range of qualities that make for a competent farmer. If possible the articles should cover some traditional types of farming; corn, rice, dairy and livestock. The articles should also contain some unfamiliar types of farming like; deer, ostrich and mushroom. The folder is homework for additional discussion at the beginning of the next class.
4) Define commercial fishing as a group and put the definition on the board.
5) Define aquaculture as a group and put the group definition on the board.
6) The extension agent should prepare a folder for each farmer with articles about what it takes to be a successful aquaculturist. Articles should contain some familiar types of aquaculture and also some quite unfamiliar types. The folder is homework for additional discussion at the beginning of the next class.
7) Homework: After reading the handouts each farmer should do a personal inventory. What qualities do they feel they possess that would make them good farmers/

(a) (b)

Figure 26. A) Oyster shells strung together and hung from rods in culture tank. B) Commercial Japanese conchocelis farm for indoor summer culture season.

(c) (d)

Figure 26. C) Hard clam shells with conchocelis filaments growing within calcium carbonate shell matrix. D) Sea scallop shells with fully matured conchocelis filaments ready to release conchospores into the seeding tank.

aquaculturists. What qualities do they need to improve on?

Aquaculture extension efforts should focus on four primary objectives: farmer recruitment; nori farming training course; training manual; and extension efforts. An eight week nori farming extension course covers biology, economics, equipment, assembly, seeding, cultivation, and harvesting. Recruitment efforts include a nori training seminar, advertisements in local newspapers, radio, and internet as well as professional publications.

Conchocelis Culture

In Asia, the culture of conchocelis starts around mid spring (March/April). As in nature, the commercial conchocelis phase is grown on oyster, clam, and scallop shells (Figure 26 a,b,c, d). The conchocelis farmer constructs greenhouses for summer cultivation of the filaments within the oyster, clam or scallop shells.

However, in Japan artificial substrata, made of transparent vinyl films covered with calcite granules, are beginning to be used as substitutes for mollusc shells. Usually, in China the natural oyster, clam or scallop or artificial shells are placed on the bottom of shallow tanks filled with seawater for seeding to take place (Figure 27)

Seeding of the cleaned shells may be accomplished by introducing chopped pieces of fertile (ripe) *Porphyra* blades into the seeding tank, which are then removed after the release of the zygotospores. The zygotospore suspension can be prepared artificially by either air drying the fertile thalli overnight and then immersing them in seawater for 4 to 5 hours the next morning (this will induce mass shedding of zygotospores) or by grinding the fertile blades and separating the suspension of zygotospores by filtration. The zygotospore suspension is then introduced into the seeding tank, where the spores settle on the calcium carbonate-based substrata. The zygotospores germinate and bore into the shells or the artificial substrates under very low light conditions. An alternative method, growing in popularity, is the mass cultivation of conchocelis that is then seeded directly on calcareous substrates. The advantage of this technique is the use of defined strains that give a more consistent crop. In each case, the seeded substrates are kept in large tanks (0.25–0.5 meter deep). The substrates are either hung vertically (Figure 26 a,b,c,d) or spread across the bottom (Figure 27). The conchocelis is allowed to develop during the rest of the summer under slightly higher light levels at 16:8 hour Light:Dark and at 23°C.

Figure 27. Indoor Chinese conchocelis shell culture tank with clam shells laid flat and side by side for several months of conchocelis growth. Chinese nori nets are laid on top of these conchocelis laden claim shells just prior to conchospore release.

Bioreactors have been developed for the laboratory culture of "free-living conchocelis" of native North American *Porphyra* species (Figure 28). Clones of conchocelis are vegetatively propagated to produce conchosporangia which in turn produces huge amounts of conchospores (Figure 29). Large amounts of conchocelis can be produced and maintained in bioreactors under controlled temperature, light, photoperiod, and salinity. The process is less cumbersome and requires less labor. Conchosporangia formation and maturation are inhibited by reducing the salinity of the seawater to 22 parts per thousand. Mass release of conchospores can be induced as needed and seeded onto nets.

Conchospore release is promoted by stirring, use of compressed air bubbling, or by treating cultures with low-temperature seawater (18° to 20°C). For the latter treatment, conchocelis-bearing shells are transferred to low-temperature seawater tanks 5 to 7 days prior to seeding. Replacing the seawater in conchocelis culture tanks and adding vitamin B_{12} can also promote conchospore release. To inhibit conchospore release, conchocelis tanks may be covered with black vinyl sheets and the culture seawater maintained calm with little, if any, aeration.

Seeding Culture Nets

Although the traditional method of the cultivation of *Porphyra* (natural seeding of conchospores on bamboo sticks and nets) may still be practiced in some areas, the bulk of nori produced in Japan, China, and Korea depends on artificial seeding of conchospores from hatchery grown conchocelis. In these countries, seeding commences in the autumn when the water temperature is 23 to 24°C. The nets are 1.8 to 2 m wide and 18 to 45 m long, made of synthetic

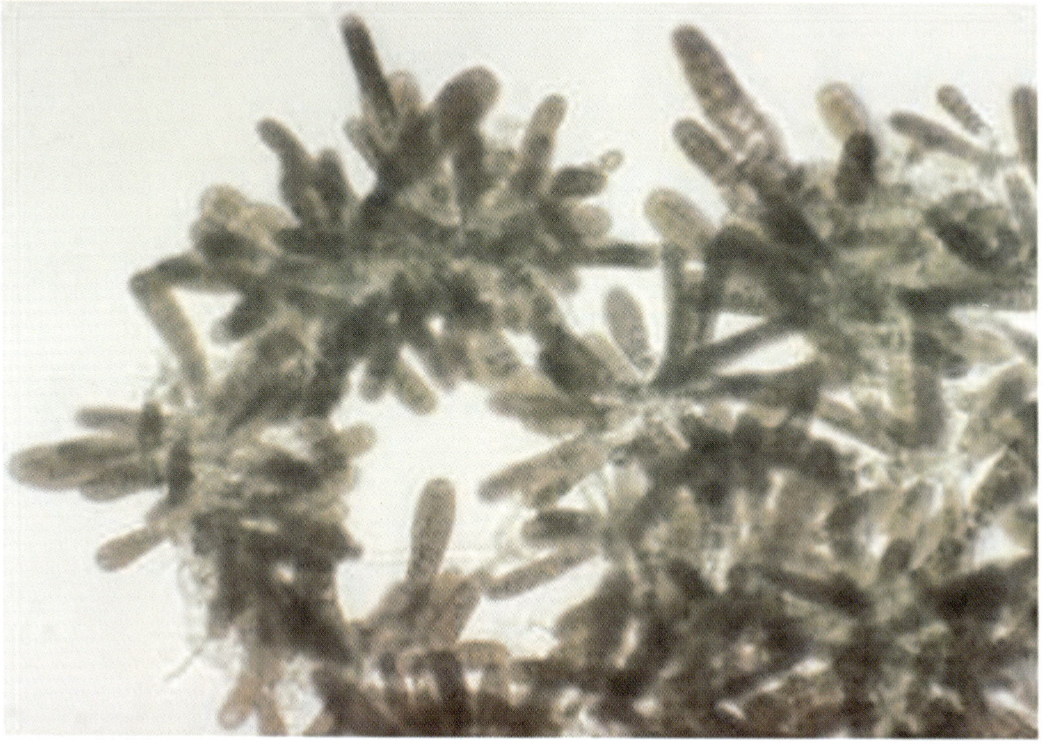

Figure 28. Shell and/or calcium carbonate-free conchosporangial culture. (Courtesy: D. Sahoo, India)

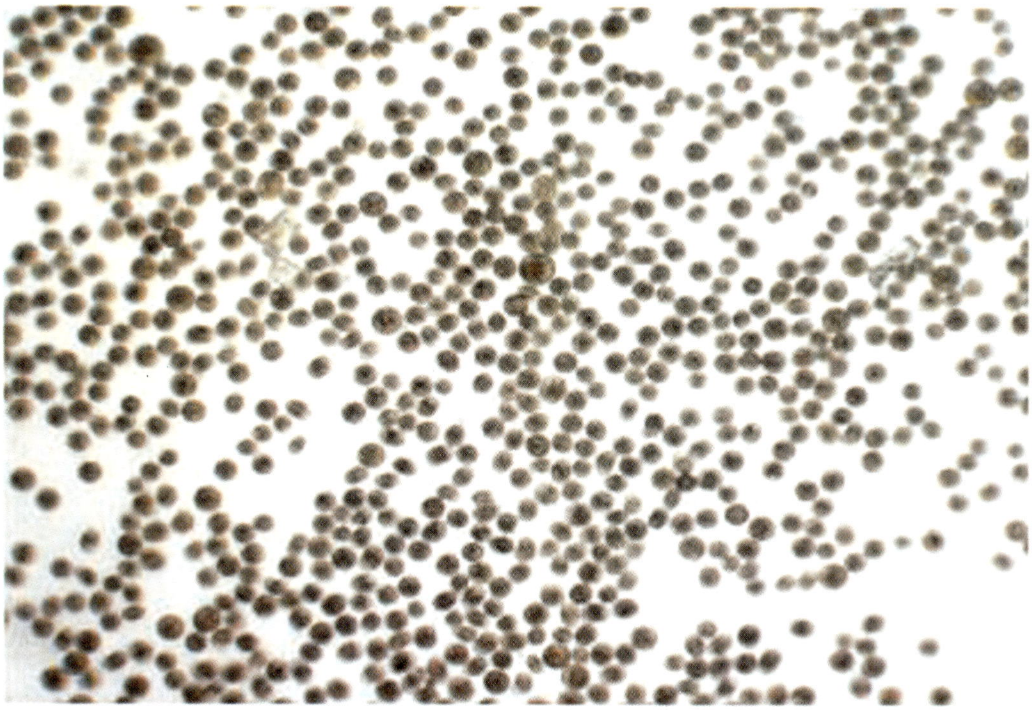

Figure 29. Mass release of conchospores. (Courtesy: D. Sahoo, India)

twine 3 to 5 mm in diameter. The nets have a mesh size ranging from 15 - 18 cm or 30 - 35 cm stretch mesh size (Figure 30).

The twine consists of three circular shaped rather than sharp-angled or square-shaped fibers; nylon for strength, polypropylene and/or polyethylene for softness, and a third fiber, which enhances the sticking of the spores. The nets have the capacity to absorb water, swell rapidly and dry very quickly. These characteristics are essential during the nursery cultivation phase, which provides the nori seedlings with a daily drying period, which kills diatoms and microbes. As the nets are raised out of the water to dry and then dropped back into the water, the net fibers shrinks and swells rapidly, causing the dead diatoms and microbes to be displaced from the net. The net's fibers are tuned to the local environments as a function of predominant weather patterns experienced, i.e. how many foggy and rainy days there are during the cultivation season.

Techniques for seeding nori nets:

1. outdoor seeding is carried out in nursery grounds by setting up layers of (12 to 16) nets on support systems. Hatchery-produced mature conchocelis on shells or on artificial substrata are placed in plastic bags and hung under the nets. The conchospores float in the water and are collected on the nets.
2. Outdoor seeding is accomplished in a stepwise manner by fixing the net either over a rotary wheel or a belt conveyor that is rotated in a seeding tank containing mature conchospores as shown in Figures 31 a,b,c,d.
3. Another method, popular in China, is the indoor seeding of nets that are suspended

Figure 30. Nori nets stacked together (6 to 12 nets at a time). These nets have just been seeded and are starting the nursery cultivation stage.

(a)

(b)

Figure 31. Nori net seed wheel with lower portion of net loaded wheel is immersed in the seed tank full of recently released conchospores. A) Japanese corporate farm workers carry fertile oyster shells to seed wheel tank. B) USA seed wheel.

(c)

(d)

Figure 31. C) Seed nets being removed from seed wheel tank D) seeded nets being carried to holding tank.

on the surface of the water in shallow seeding tanks (Figure 32 a & b). The conchospores are inoculated by circulating water via a submersible pump that is in the tank.

The nets that are seeded with the conchospores are stacked into bundles of 4-12 nets. These stacks of nets are transferred to the sea for nursery

Figure 32. A) Indoor Chinese conchocelis shell culture facility; B) Individual, shallow shell culture tank with clam shells laid flat and side by side for several months of conchocelis growth. Chinese nori nets are laid on top of these conchocelis laden claim shells just prior to conchospore release.

cultivation. In nursery culture, the nets are put into the sea and carefully monitored for blade development.

Nursery Sporeling Culture

During this early nursery sporeling culture stage, the nets are placed into a nursery system, which is designed to raise the nets out of the water for daily drying, simulating intertidal variations exposing the young thalli to air and sun. This exposure is necessary to inhibit fouling organisms (e.g., other seaweed species or diatoms). In Japan, a popular floating system, called the "Urano Fujio Ikada" system, is used for this process (Figure 33 a & b).

In China, Professor X. G. Fei has developed

Figure 33. Urano Fujio Ikada system. A gift of Japanese nori farmer, Urano. The technology allows for one person to raise a nori net in < 15 seconds. A) Immersed nets; B) Emersed nets

a floating A-frame system (Figure 34a) that can be easily raised and lowered for supporting nursery nets and controlling the degree of exposure. A Korean design utilized in low energy environments, amazing in its simplicity, attaches net frames to Styrofoam floats. The net frame float attachments will keep nets on the surface but if the net is rotated 180 degrees, the Styrofoam floats now orientate the nets out of the water for the drying phase of the nursery stage (Figure 34b).

Additionally, the pole culture system acts as

Figure 34. A. Floating "A-frame" system which allows for easy conversion to exposed net condition. B. Korean Styrofoam Ikada/production system.

both the nursery and production system facility as one can control the net emersion/immersion process (Figure 35).

After 20 - 30 day nursery phase, where the blades are 1 to 3 cm in length, the nets can be transferred to the farm sites or frozen for later use. The growth of *Porphyra* thalli is inhibited mostly by high temperatures during mid to late November or as a result of disease or lack of nutrients. In these cases, nori production yields and sheet quality can be significantly reduced. To avoid this, the technique of frozen storage of nori nets was devised in 1969. These nets are initially air dried to reduce the water content of *Porphyra* to 20 to 40%, packed in polyethylene bags and then stored at –20°C. The seedlings that are stored under such conditions remain alive for 6 months and can be used whenever required. The frozen nets can then be used to replace tired, lost or damaged ones.

Production System

- What type of culture technique Pole/Floating / Semi-floating
- Choosing a culture site Pole/Floating/Semi-floating
- Equipment and labor requirements Pole/ Floating/Semi-floating

Figure 35. Fixed pole farm with nets emersed out of the water.

OBJECTIVES: The farmer will understand the three forms of nori production culture: Pole (Figure 36 a & b), Floating (Figure 37 a & b), and Semi-Floating Systems (Figure 38 a, b, c).

The farmer will understand the pros and cons for each culture technique. The farmer will be able to make an informed decision as to the type of culture that they will establish.

Figure 36. Fixed Pole Farms A) metal poles, USA; B) bamboo and plastic composite, Japan (Courtesy: Professor Masao Ohno)

Figure 37. A) Aerial of open water floating production system – USA. B) Individual nori net from open water floating system.

Figure 38. A. Semi-floating nori production system developed by Professor Fei of Academica Sinica, Qindao, China with D. Sahoo in 2003. B) Semi floating system nori net's emersed.

Figure 38. C) Ideal for low energy, intertidal zones.

RATIONALE: The farmer will have to choose a culture technique and needs appropriate and accurate information to base their decision on.

READING MATERIALS AND ACTIVITIES:

1) A brief description of the pole *shichu*, floating *furyu*, and semi-floating culture systems. Open water floating and semi-floating culture is done only in subtidal areas and all moorings consist of anchors. Pole culture is shallow water or intertidal culture, where the nets are tied to fixed poles. (Handouts and overheads showing all culture techniques.)

Floating open water nori cultivation system dimensions require large tracts of the river, bay, or ocean. Large areas are required, 0.20 to 0.25 km^2 for a commercial farm, as nets must be widely spaced to allow adequate water flow through the entire farm (Egan, 1990). The total lease site for floating cultivation system should be six times the area taken up by the nori nets themselves (Miura, 1975). The total footprint of a pole culture site is significantly reduced as you do not need the extended anchorage infrastructure to maintain system integrity. Lease size requirements have resulted in water use conflicts with recreational and commercial fisheries, recreational and commercial boaters, waterfront land owners, the coast guard, and many other stakeholders. As a result, the acquisition of permits and leases has become a very difficult and lengthy process, and is considered to be the single greatest barrier to the expansion of seaweed industries (Mumford, 1990; Levine, 2006).

Advantages to open water floating culture:

- The lease site is not limited to shallow (<6 meter) water depth
- Increased net accessibility for harvest, treatment and cleaning
- Reduced capital expenditure
- Installation goes quickly

Advantages to pole culture:

- Decreased storm and strong tide damage
- System integrity is easier to maintain
- Decrease fouling from floating debris
- The nori is usually of higher quality due to closer correlation to natural growing conditions realized in the intertidal zone

Disadvantages to floating culture:

- Full exposure to storms and strong tides
- System integrity needs constant monitoring and tuning
- Fouling from floating debris
- Not the natural, intertidal growing conditions for nori

Disadvantages to pole culture:

- Lease site is limited to shallow (<6 meters) water depth
- Decreased net accessibility for harvest, treatment and cleaning
- Heavy capital expenditure in the beginning
- Lengthy assembly.

2) Ask farmers what additional pros and cons they can see with each culture technique. A discussion of the yin:yang relationship between the two types of culture may illicit more responses. A simple statement may also further discussion, such as, "Everything in aquaculture is a plus and a minus at the same time". Pole culture while being less susceptible to storms and strong tides has decreased accessibility to nets for care and maintenance. The successful aquaculturist is able to maximize the advantages and minimize the challenges.
3) The above discussion is a good transition to the next step on lease site selection if the discussion is progressing in this direction; it may be beneficial to let it continue until it runs its course.

Choosing a Culture Site Pole/Floating

OBJECTIVES: The farmer will understand the variables to consider for site selection. The farmer will be able to progress through a logical process of site selection. The farmer will be able to determine appropriate types of culture for various sites and give the pros and cons for the sites selected.

RATIONALE: Inappropriate site selection will cause undue hardship, decreases productivity, and increases the possibility of failure. Site selection will be assessed through the following areas: biological, culture, and traditional fisheries.

READING MATERIALS AND ACTIVITIES:

1) Several site scenarios will be presented to farmers for them to assess in class.
2) Farmers will develop an ideal site for pole and floating culture and discuss it why they believe this is the ideal site scenario.

PRESENTATION: Site selection criteria include weather, site energy, water depth, wind and wave exposure, water quality, nutrient levels and proximity to boat moorings. Successful site selection is a function maximizing each of these parameters as a function of the whole. Optimal criteria include: high levels of fog, high energy with rapid currents, water depths at MLW of ~ 13 meters, and minimum exposure to wind and waves. Avoid land based pollution, shipping lanes, areas with heavy commercial fishing or recreational boat traffic, and sea grass beds. Indications of appropriate cultivation areas include: presence of wild native nori, and the abundance of annual macroalgae (i.e. *Enteromorpha* and *Ulva*).

The individual site's specific environmental parameters will determine the most efficient system design. There are virtually infinite number of designs to deploy. Current, water depth, seaweed and debris abundance, wind exposure, lease site dimensions and orientation, distance from land, other leases, moorings, docks, piers, and channels are just a few of the factors that determine the system configuration. Minimum biological requirements for either pole or floating culture must be met before any other considerations for site selection. A well-protected site with no nutrients will only allow for peaceful failure. Trial and error is the only way to learn the best design for your lease site.

Low energy sites with less than one knot of current need in excess of 100 mg/L of nitrogen to be considered for commercial production. Recent studies have determined that effluent from fish farms will support 35 nets per metric ton of fish per year. 150,000 fish farm will produce nutrients to support over 10,000 nets per year. While no one is proposing trying to manage 10,000 nets, it is important to recognize this untapped resource. Light levels must also meet the requirements for the species grown. This will vary as to which season you are growing in and may require nets to be stationed at different levels in the water column at different times of the year. Seasonal turbidity will affect light levels and should also be taken into account. The relationship between available nutrients, light levels and product quality is very strong. To this end, excess nutrients are never a problem and excess light is always a problem. Temperature is another biological parameter, which must be in sync with the species grown. Once basic biological parameters are met one can then decide which type of culture is best suited for an individual environment.

Both floating culture and pole culture require soft bottoms for either driving poles or setting anchors. Pole culture requires a shallow water environment and floating culture does not. Both systems should have adequate flow to move nutrients through the system, in excess of one knot. A protected area is more ideal than one that is not. One should base this decision on what seasons they plan to operate and determine the prevailing wind for that time of year. To work the culture ground one needs less than one-foot seas, otherwise the boats will cause damage to the nets and system. For a floating system everything is in the water and growers work off the bows of low boats, face down. While it is possible to work, and sometimes necessary in higher seas it is counter productive and usually not very enjoyable. Getting to the site should not be overlooked. A well-protected site may not be accessible during certain months and it is best to determine this prior to placing your system there.

A third area to consider in site selection is what traditional fisheries, if any exist. While this may or may not come to light during a lease application process one wants to determine what use the area currently undergoes. Fishing, shipping lanes, and pleasure boating water utilization should all be determined: preventing future water use conflicts.

Equipment and Labor Requirements Pole/ Floating

OBJECTIVE: The farmer will be made aware of the supplies required per 100 saku unit of Pole/ Floating production system. The farmer will understand the very different labor and equipment requirements for Pole/Floating production.

RATIONALE: Each individual has has own preferences. An individual will be more successful working with something they feel comfortable with.

READING MATERIALS AND ACTIVITIES

1) Pictures of 100 saku production systems of both pole and floating culture.
2) Equipment lists compiled with farmers for each type of system after initial presentation.
3) Each farmer is to decide which system they prefer and why as homework. Use for class discussion at the beginning of the next class. At the end of the course this question should be asked again and the answers and reasons for preference compared with those given initially.

PRESENTATION: Pole culture is initially more labor and capital intensive. Eight to twelve poles per net need to be driven into the bottom. That is 800 to 1200 poles per 100-saku unit. Poles are driven five to ten feet into the bottom and 10 to 15 feet of the poles should remain. Poles then need to be 15 to 25 feet long depending on local tidal amplitudes. Poles in high-energy areas should be driven farther into the bottom. Poles, depending on material, source and volume can cost five to twenty dollars each. Poles can be driven by hand, but a pole driver is recommended to decrease the time, energy and injury from building the system. Poles can only be driven around low water, which may be good conditioning since all activities for pole culture must be planned around the tides.

Floating culture is less labor intensive at the beginning and more accessible always. Anchors for the system are lightweight, ~ 32 kg and easy to build (Figure 39).

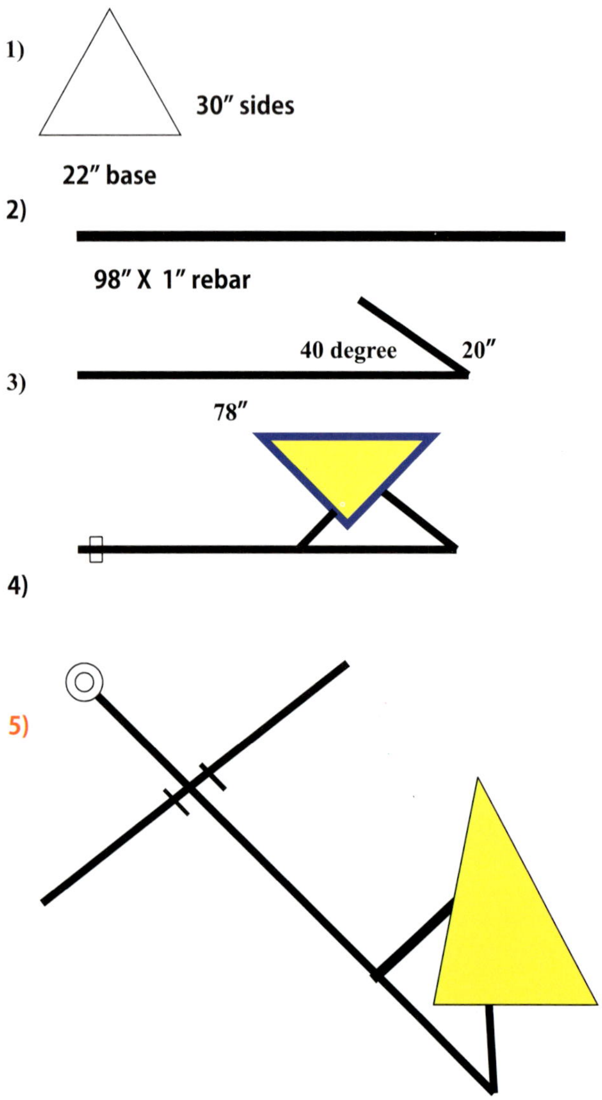

Figure 39. Anchoring system design.

The rest of the system is a grid made of ropes into which nets are tied. All lines are fabricated on land. The system can be placed in the water in 2-3 days and tightened at low water. All anchor lines are tied into system lines at a loop. A sheet bend is used on the loop. All tuning of the system is to occur at these points. Resetting anchors should only take place if anchors have broken free. Never tune the system by resetting anchors.

System Construction

OBJECTIVE: The farmer will be able to perform all pre-assembly work for a 100-saku floating and a 10-saku pole production system.

RATIONALE: The more familiar the farmers are with the two systems the better prepared they will be when they make their decision as to which type of culture system they wish to install.

READING MATERIAL AND ACTIVITIES:

1) Farmers will do all onshore preparation for a 10-saku-pole system.
2) Farmers will do all onshore preparation for a 100-saku floating system.

PRESENTATION: Onshore preparation for a pole culture system is minimal. Having acquired poles, place them as close to the boat departure-site as possible. Loads will be limited by boat size. While small boats are needed to work within both systems at least one large workboat is needed to convey material and personnel to and from the site. A manual nori pole driver can be made quickly and cheaply (Figure 40).

Six-foot lines of 5/16″ rope should be prepared at this time. Three lines per pole to allow for height adjustment. Rope qualities should include; UV resistance, remain soft and pliable, strong and affordable. A number of manufacturers are producing a green poly-steel type rope with all of these qualities.

Onshore preparation for a floating system is more involved, but takes much less time to install.

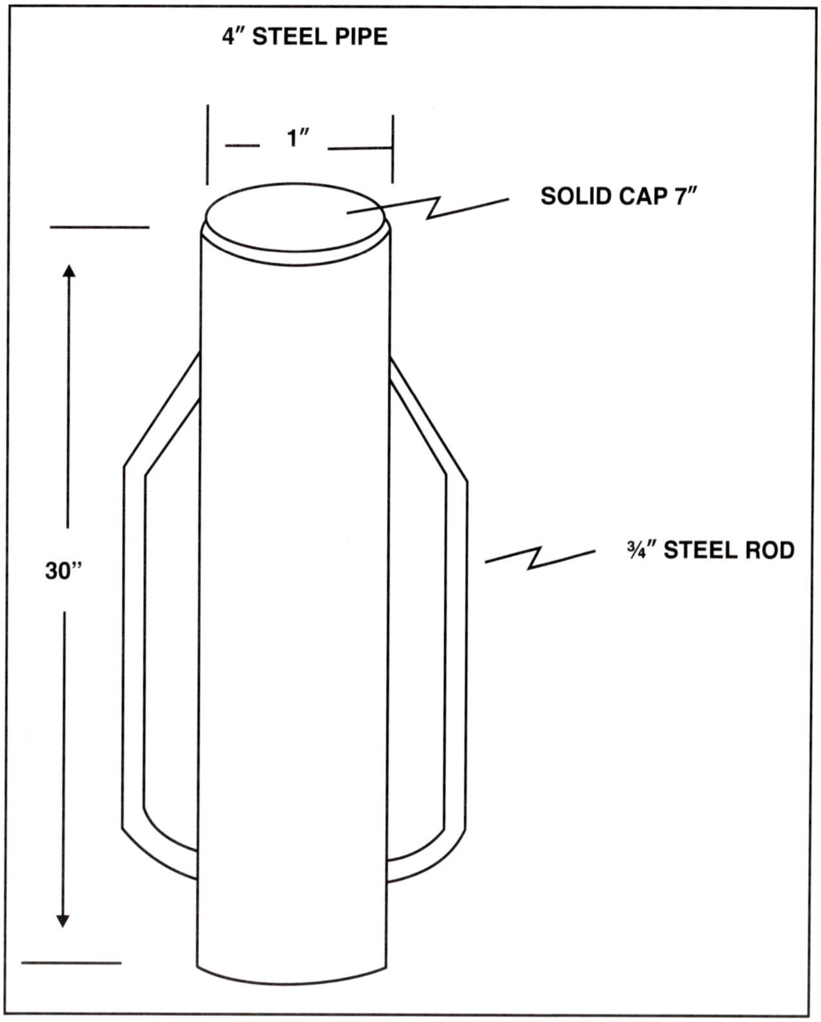

Figure 40. Pole driver used to set fixed poles.

A farmer will need the following for each 100-saku system that you wish to install.

1) 44 anchors with cross bars and pins.
2) 44 × ¾″ anchor lines with a scope of 5 to 1.
3) 44 × ¾″ tag lines with floats.
4) 44 black ball mussel floats
5) 20–20′ yokes with one large center loop.
6) 200-70′ × 5/16″ production lines with 5-7 evenly spaced 3″ floats.
7) 2–700′ × ¾″ side lines with end loops, mark the line with tape every 70′ for installation of middle lines.
8) 11–140′ × 3/4″ middle and end lines with end loops. Two of these are end lines which should have single loops from 5/16″ rope tied into the rope every 7″. The remaining 9 lines are middle lines and they need double loops of 5/16″ rope tied through the middle line rope every 7′.

MOORINGS

Nori culture grounds provide a variety of bottoms and current flows causing a profound challenge for the nori grower to secure his site. No one mooring system can be described that will adequately suit all nori farm sites. The author has used 3,000 lb. granite blocks, 30' pilings, 3 foot 1-1/2" steel pins drilled into ledges, standard 70-lb nori fluke anchors (Figure 41), 80-lb Danforth anchors, 3 foot × 2" pipe anchors (Figure 42), and 600 lb anchors made of railroad rails with varying

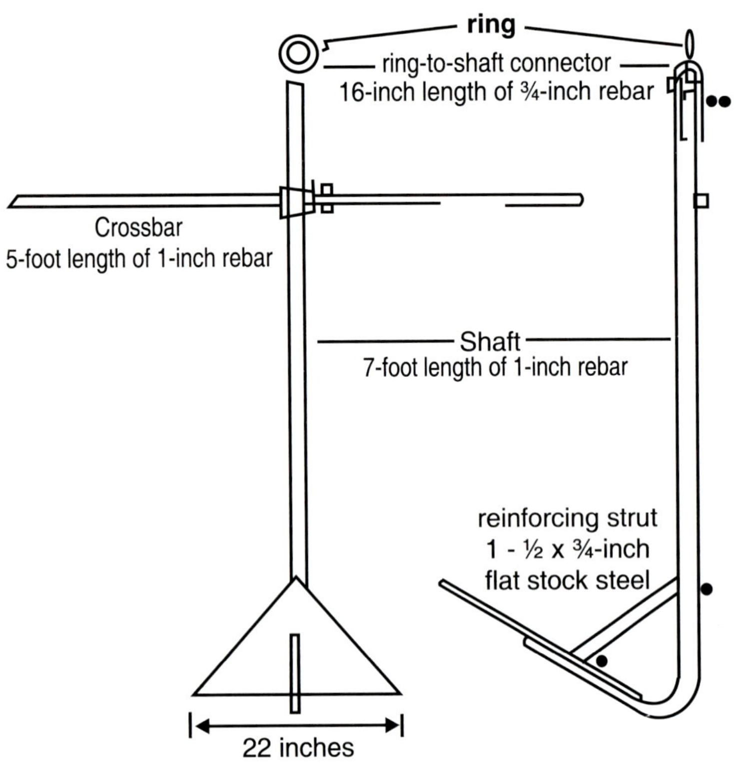

Figure 41. Fully constructed fluke anchor to set the open water floating systems.

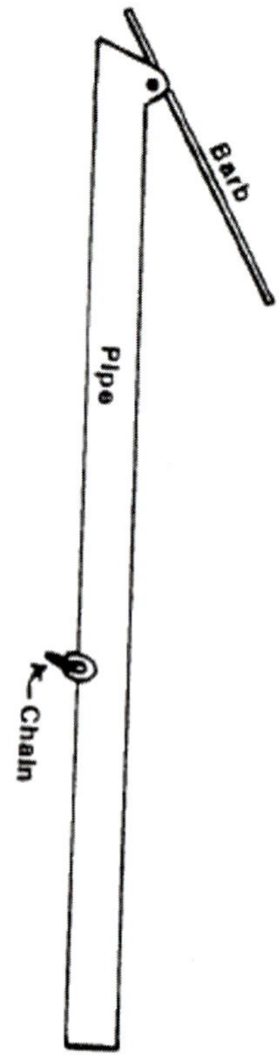

Figure 42. Pipe anchor used in low energy environments to secure open water floating systems.

degrees of success and numerous failures. The system must be securely anchored to have any success, and the cost of the mooring will be the restrictive factor. These anchors work well in mud or sand bottoms in maximum currents of less than 2 knots. Maximum current is determined at the strongest spring tide during the growing season. A mooring system is worthless if it holds well during all tides except the 2-3 days of the strongest spring tide and the system collapses.

Rail anchors (Figure 43) can be made rather cheaply because used rails sell for $.10/lb at a junkyard. One welder can construct two, 600-lb rail anchors per day. They will fit on the back of a pickup truck, and a commercial boat should be hired to put them in place. This may cost as much as the construction of the anchor.

Three thousand pound granite moorings have been used to secure small nori net systems. The bottom must be mud or sand. If the bottom is ledge or cobble, these moorings move with incredible ease and are worthless. Do not consider making moorings out of concrete. Concrete does not weigh much more than water and will not work as a nori system mooring but works very well as a tension anchor for system tuning (Figure 44).

Three-foot pipe anchors (Figure 42) have been used successfully in Japan and U.S.A. The bottom must be mud or sand. Clay substrate is nearly impossible to work with. Attempts to use high-pressure water hose for placement met with total failure. The waterjet created a large hole in the clay bottom, which did not fill in around the pipe anchor.

1) The fluke anchor (Figure 39 & 41) is a triangle with a 22″ base and 30″ sides cut from ¼″ plate steel.
2) The neck of the anchor is 1″ rebar. Cut one 98″ length of the 1″ rebar.
3) At 78″ heat and bend the rebar to a 40-degree angle.

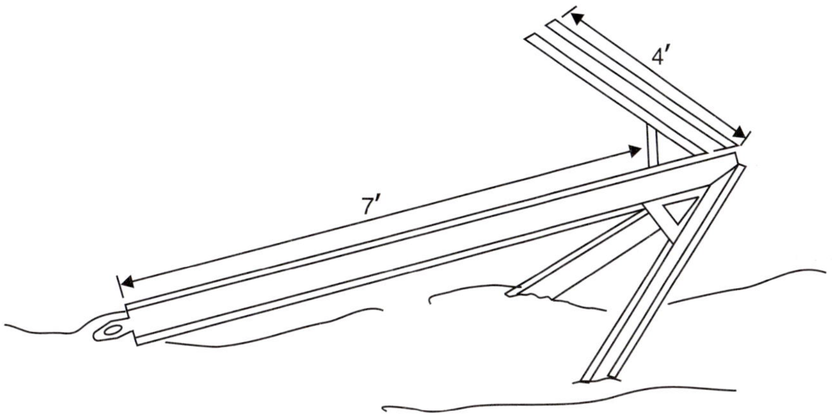

Figure 43. Fully constructed rail anchor, utilized in open water floating systems situated in fast moving waters.

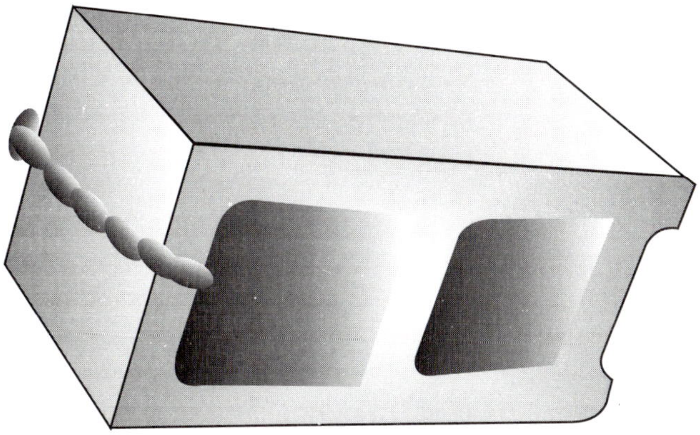

Figure 44. Standard concrete cinder block harnessed as a tension anchor to "tune" and tighten the open water floating systems.

4) Cut a piece of 1″ by ¼″ flat stock 12″ long, this is the brace or gusset between the bent rebar.
5) Weld the fluke onto the rebar. Eight inches of the rebar should be in contact with the fluke.
6) Weld the brace to the rebar neck at 66″. Now weld the other end of the brace to the fluke and rebar.
7) 30″ from the top of the neck it is necessary to weld a collar for the cross bar. Weld a 3″ long by ¼″ thick section of pipe cross wise to the neck, the cross bar will pass through this.
8) Weld a 4″ ring to the top of the neck, this is to tie the anchor line too.
9) The cross bar is a 5′ piece of the 1″ rebar with a stay welded two inches off center. Two inches off center in the other direction drill a hole large enough to accommodate your anchor pins (.172″ works well for pin size).

Ropes and Knots

There are two important considerations concerning ropes used in nori farming; nori ropes should float and they should be able to hold a knot. While loose floating ropes in the culture area may foul outboard motor propellers, the ability to find dropped ropes or the item attached to them quickly outweighs this disadvantage. Wall brand "steel rope", a blend of several fibers, is ideal, being inexpensive, easy to tie, floats, and resists chafing. Mooring and system frame lines should be 3/4″, production lines and lines for tying nets in can be 5/16 or 3/8″.

Three types of knots are utilized in nori cultivation systems. The sheet bend with a loop (Figure 45a) is used to tie mooring and production lines to the system and for tying nori nets to the production lines. These three locations require frequent adjusting and this knot allows efficient untying by simply pulling the loop through. If there is a lot of extra rope after tightening mooring lines, it can be coiled, placed in the loop, and the loop tightened to prevent the extra rope from uncoiling and being a nuisance. When tying a net to the production line always begin the knot by running the end from the top of the net down (Figure 45b), never from the bottom up. When the knot is completed, the loop will be on top of the net. The advantage of this becomes obvious

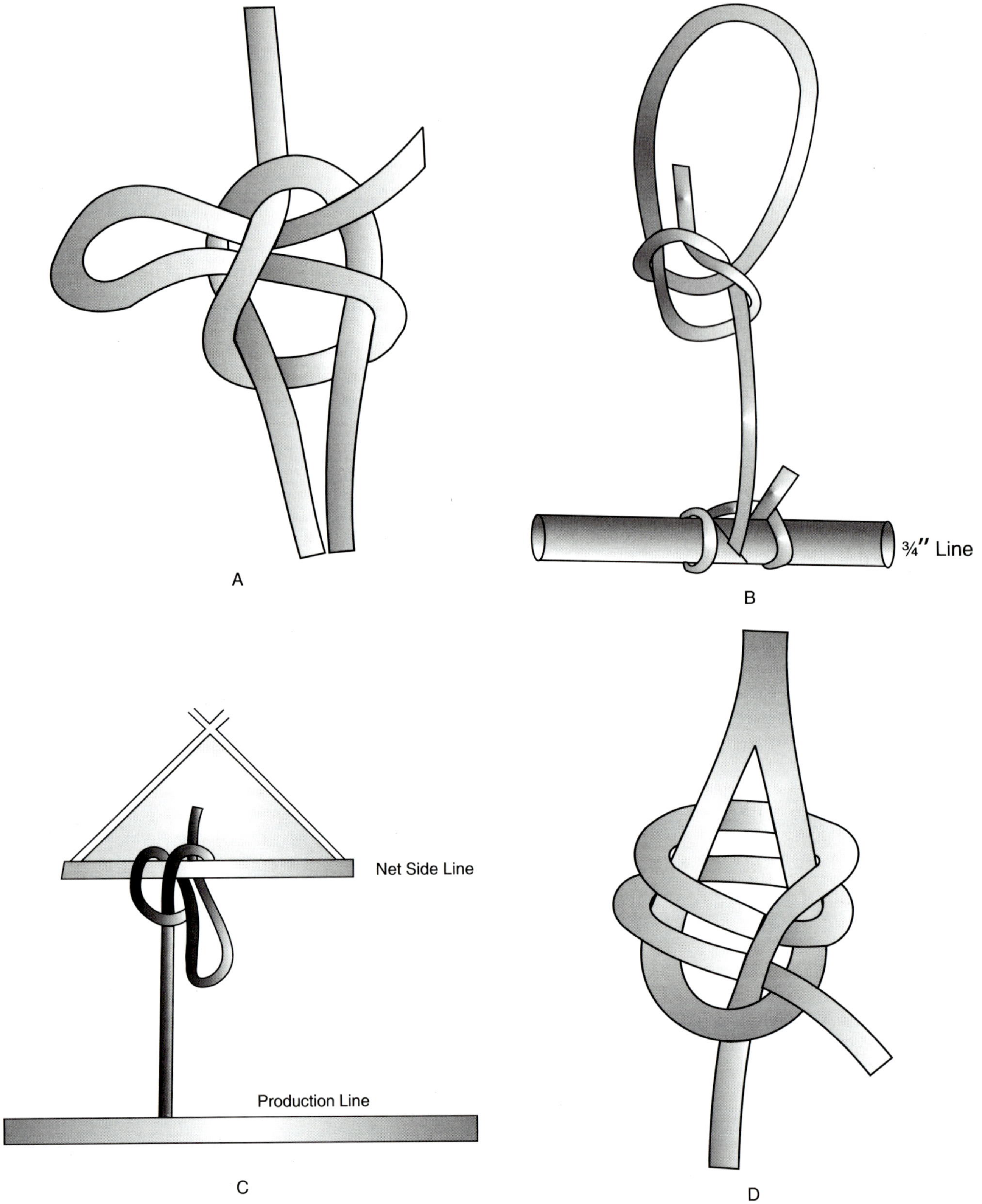

Figure 45. A. Sheet bend with a loop knot. B. Sheet bend with a loop knot used to secure a net to a production line. C. A bowling knot used, both secure and easy to untie. D. Double sheet bend knot used only to secure anchorage.

when harvesting the net, as the loop stays out of the harvest blade. The bowline (Figure 45c) is used to tie net side lines to production lines, loops in the end of system lines, and boats to systems. The bowline is a very secure knot and allows untying no matter how tight or how long it has been in position. This knot can also be used to tie buoys, anchor lines, and tag lines to anchors. The third knot is a double sheet bend (Figure 45D); used only to tie anchor lines to anchors. This very secure knot has the advantage over a bowline in that it can be tied with mittens on and can be tied very quickly. Once it is tightened it usually must be cut off.

Regardless of the moorings used, the 3/4" mooring line must be placed with at least a 5:1 scope, or 5 times more than the depth at high water. These lines must be tight at all times and are tightened at low water. With variable tidal ranges up to 7 meters, as the tide raises the system there must be enough stretch in the ropes to avoid pulling the nori system under water at high tide.

System Installation

OBJECTIVE: To provide hands on experience for the farmer. The farmer will be able to install both floating and pole systems.

RATIONALE: Experience makes the best teacher. Installation of the systems will provide perspective as to what the farmer is undertaking as a nori farmer, reality check.

READING MATERIALS AND ACTIVITIES:

1) The farmers will install a 10-saku-pole system.
2) The farmers will install a 100-saku floating system.

The following instructions describe a 100-saku site, 25 nets wide and 4 nets deep with 6 feet between each net row (Figures 46). The maximum current flow is less than 2 knots and the maximum depth is 30 feet. This allows the use of fluke anchors.

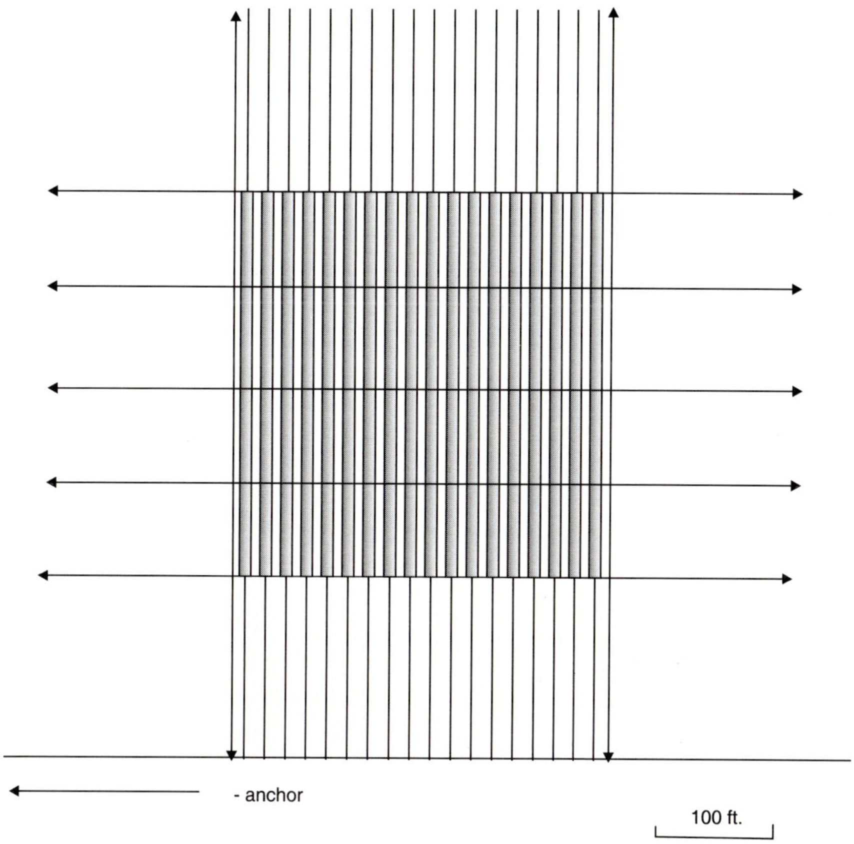

Figure 46. Diagrammatic layout of a 100-saku (net space) open water floating production system.

PRESENTATION: Orient the system so that the nets lie in the same direction as the predominant direction of the current. This can be surprisingly difficult to determine in a very high-energy river, bays and inlets.

Pole system installation is very straightforward. Start at a corner of the site. Poles are placed in pairs seven feet apart (Figure 47). The distance between pairs of poles depends on the number of pairs used. The first and last pair needs to be 70′ apart (Figure 48).

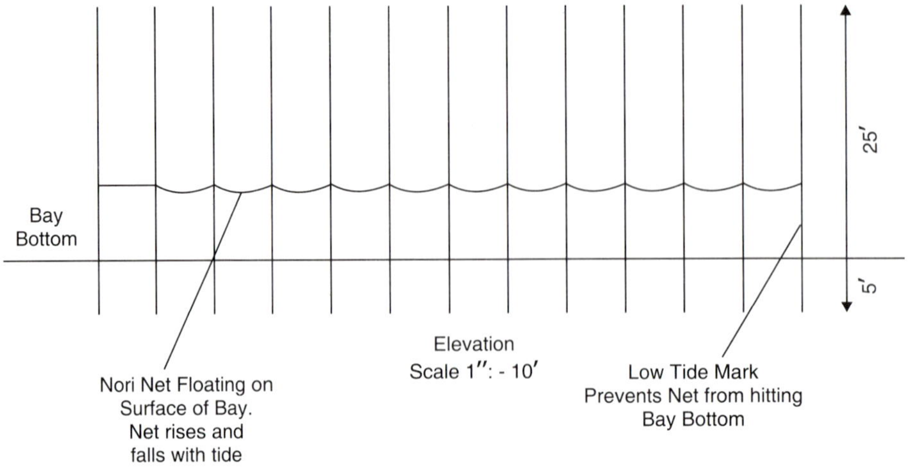

Figure 47. Diagrammatic side-view configuration of a single net within in fixed pole production system.

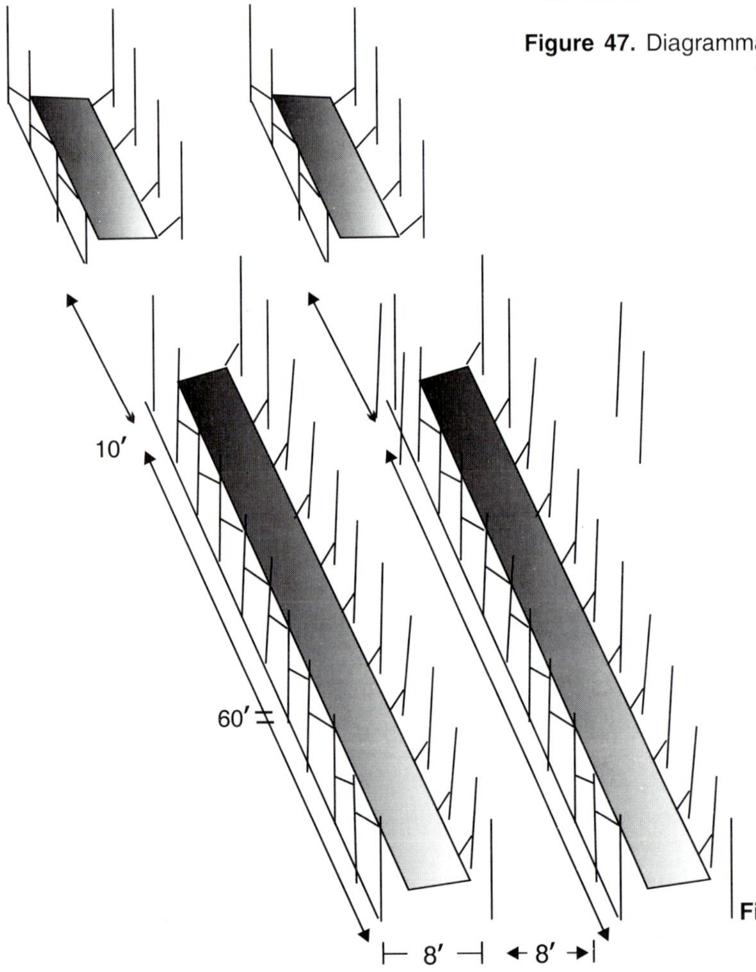

Figure 48. Fixed pole net system.

The lines on the first and last pair of poles have loop ends for tying the net rope ends into. All other lines are straight ends for tying the nets in with a sheet bend. Always maintain 7′ across between rows of nets for water flow and maneuverability. It may be helpful to place a pole and then run a line for several hundred feet and drive a second pole and use the tightened line as a guide or straight line. You can mark the line at appropriate intervals for placing poles to quicken the installation.

Floating system installation is also very straightforward.

End lines. Using 3/4" rope, measure out a length of 300 ft. Tie a bowline at each end. Five feet from an end loop, use a four-foot length of 5/16" rope to tie a production line loop into the end line (Figure 45c). Because a nori net is 5 ft. wide, tie another loop 5 feet from the first, and then the next loop will have a space of 6 ft to allow a 6' space between nets. Alternate the spacing 24 more times so that there is space for 25 nets, with an interval of 6 ft. between them. Two end lines are needed.

Middle lines. These are the same dimensions as the end lines, except there must be production line loops on both sides of the middle line, so the 5/16" lines must be 8 ft., with loops on both ends (Figure 49). Three middle lines are used.

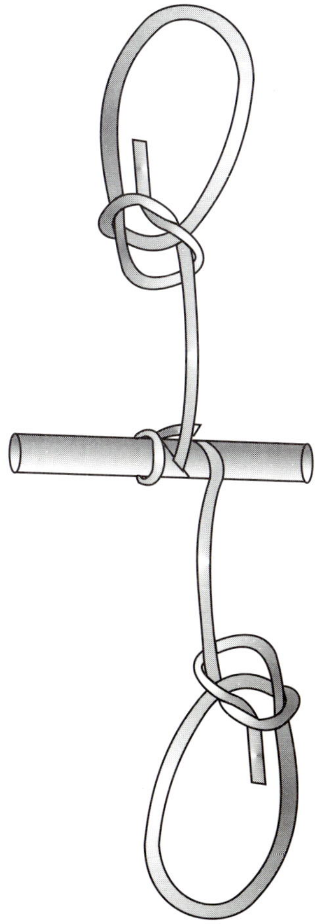

Figure 49. Middle line duel production loop ties.

Side lines. Use 3/4" rope and allow for 80 ft. per net. Because there are four rows of nets, this equals 320 ft. Tie loops at each end. At each 80-ft. segment, tie in an 8-ft. length of 3/4" rope with loops at both ends, to secure the middle lines and one mooring line. Make two sidelines. (Figure 50)

Production lines. Each production line is used to tie in one side of the net. Use 80 ft. of 5/16" line. At 7-ft. intervals tie in a 4" x 2" gill net torpedo float. This will result in 10 floats per production line. Tie a loop 3' from each end of the line for the nets to be tied in. At every float a 2' piece of 5/16" line is tied in so that the sides of the net can be attached. Two production lines are needed for each net, therefore 200 production lines must be made. (Figure 51)

Yokes. These are placed at the end of each net row to secure the mooring line. Use 20' of 3/4" rope and tie a loop in the middle of this line. The ends are secured to the end line with a sheet bend (Figure 52)

System Deployment

To avoid tangled ropes, lay the rope in a basket, weaving it back and forth until the entire rope is in the basket. If the basket is needed for another use, carefully dump the rope on the floor or deck. If this pile of rope needs to be transported, it can be picked up and placed back into a basket. At the time the rope is needed, grab the end on top of the pile, attach it to a fixed object, and unravel the rope. It will very rarely tangle.

1. Orient the system so that the nets lie in the same direction as the predominant direction of the current.

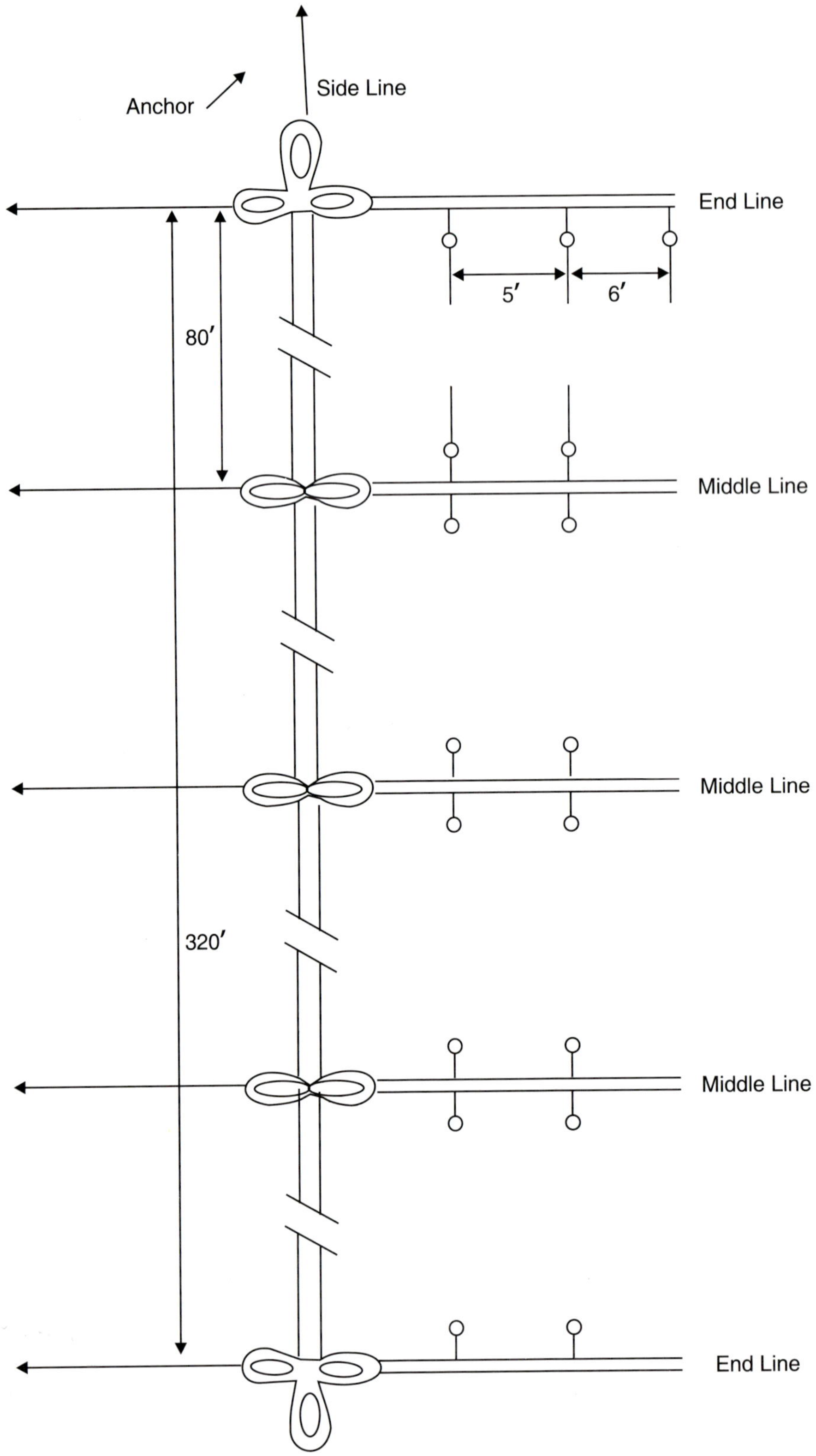

Figure 50. Sideline construction diagram.

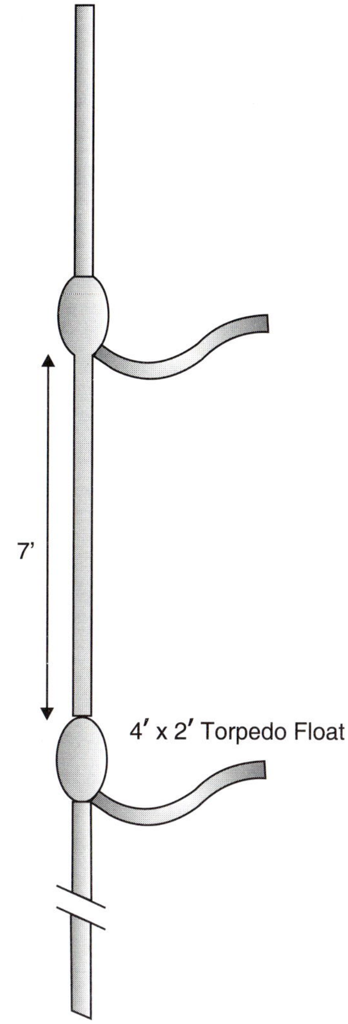

Figure 51. Production lines with torpedo float installation.

2. Tie an anchor line to an anchor using a double sheet bend knot. Tie a 30 ft. tag line to the shank of the anchor, and a floating buoy to the other end of the tag line, and drop the anchor to the bottom (Figure 53). Refer to Figure 54 for system configuration for sections 3-20.
3. Deploy M-1, the first anchor line, and run with the current to the end of the 150-ft. anchor line.
4. Tie the anchor line to S-1, a system sideline, using a sheet bend with a loop. with 10 ft extra anchor line pulled through the sideline end loop. It is very important to have at least 10 ft. of extra anchor line pulled through at this point.
5. Tie a 70-lb. float to the S-1 next to the anchor line knot, using a bowline.
6. Unravel S-1, running with the current. tie a 70-lb. buoy just in front of the loop, attach an anchor line with a sheet bend with a loop with at least 10 ft. of anchor line pulled through the side line loop.
7. Deploy M-2, tying an anchor and tag line to the anchor as in step 2.
8. Attach M-3 to the sideline loop, and deploy M-3 exactly 90 degrees to the sideline. Use the boat compass. Be sure to have at least 10 ft. extra anchor line through the side line loop. Attach the anchor and tag line as in Step 2.

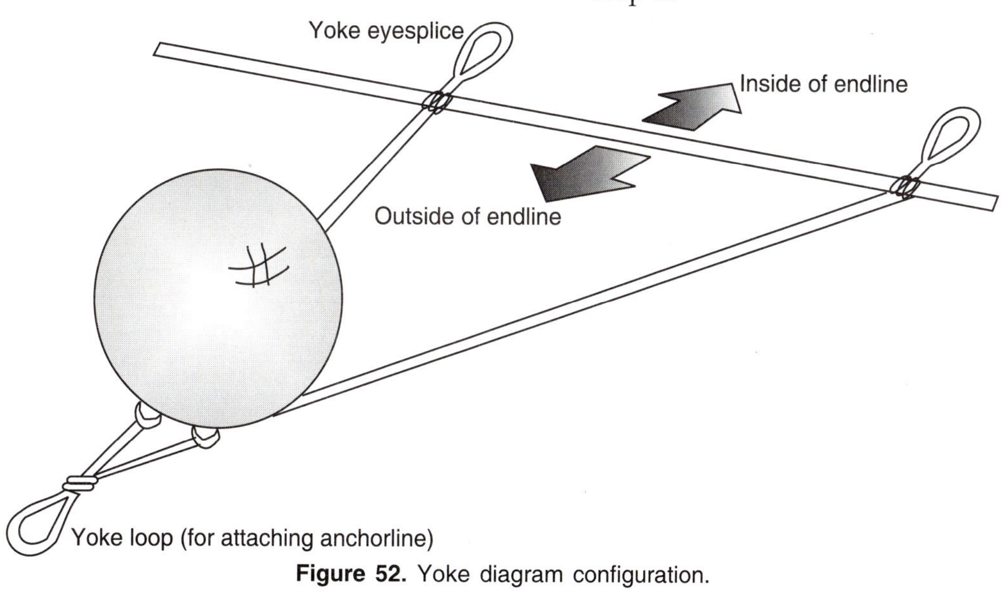

Figure 52. Yoke diagram configuration.

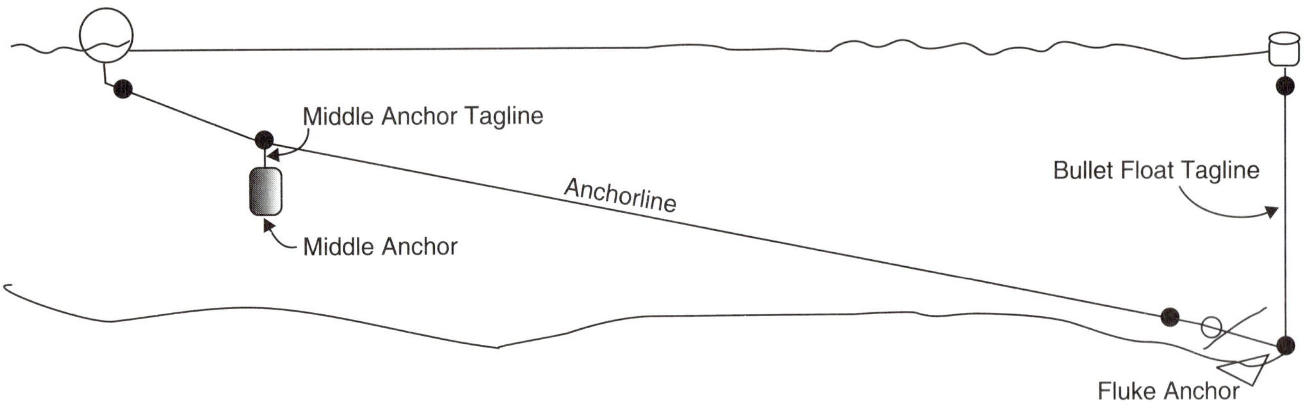

Figure 53. Anchorage configuration including fluke and tension anchors.

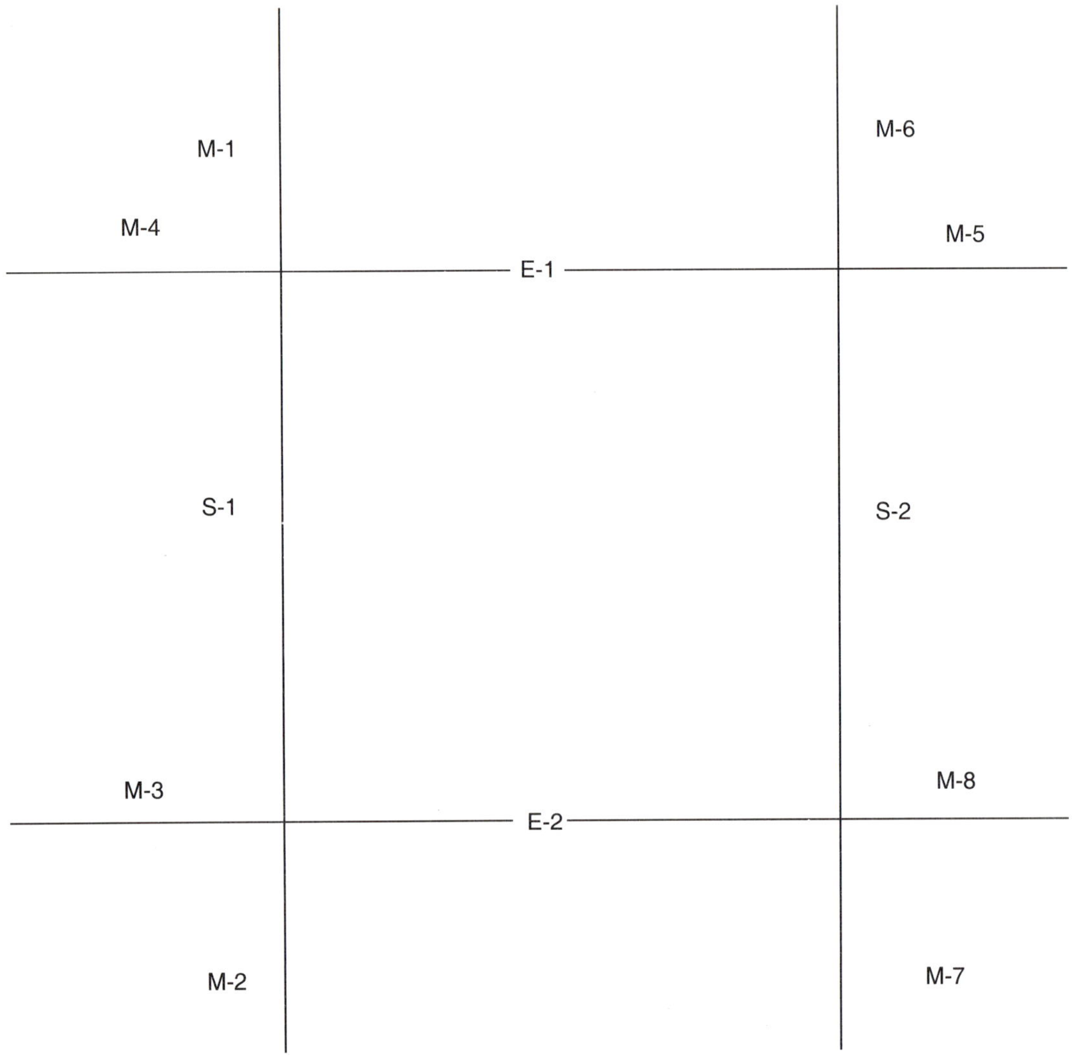

Figure 54. Open water floating production system configuration.

9. Attach M-4 and deploy in the same manner as M-3. Be careful not to pull M-3 or M-4 tight, as this will pull S-1 out of place.
10. Attach E-1, an end line, to the sideline loop using a bowline. Deploy E-1 in the same plane as M-4. Attach a 70-lb buoy just in front of the E-1 loop, tie M-5 to the loop, and deploy M-5 in the same plane. This mooring line can be tightened slightly.
11. M-6 mooring line is then attached to E-1 and run 90 degrees to E-1. Always be sure there is at least 10 ft of extra mooring line run through the loop.
12. S-2, the second side lines, can now be attached and run in the same plane as M-6. Tie a 70 -lb. buoy just in front of the loop, and attach M-7 with 10 ft extra mooring line through the loop and continue in the same plane.
13. Attach E-2, the other end line, at the corner where M-3 and S-1 are tied, and deploy in the same plane as M-3. Tie to the corner at S-2 and M-7.
14. Attach M-8 to E-2 and deploy in the same direction. This mooring line can be pulled with some tension. The system should now look like Figure 55.
15. Tie the boat at each corner and tighten the corner mooring lines, using the compass to be sure that all mooring lines at 90 degree angles to each other and the system lines are at 90 degree angles. The system must be "all-square" at this time.

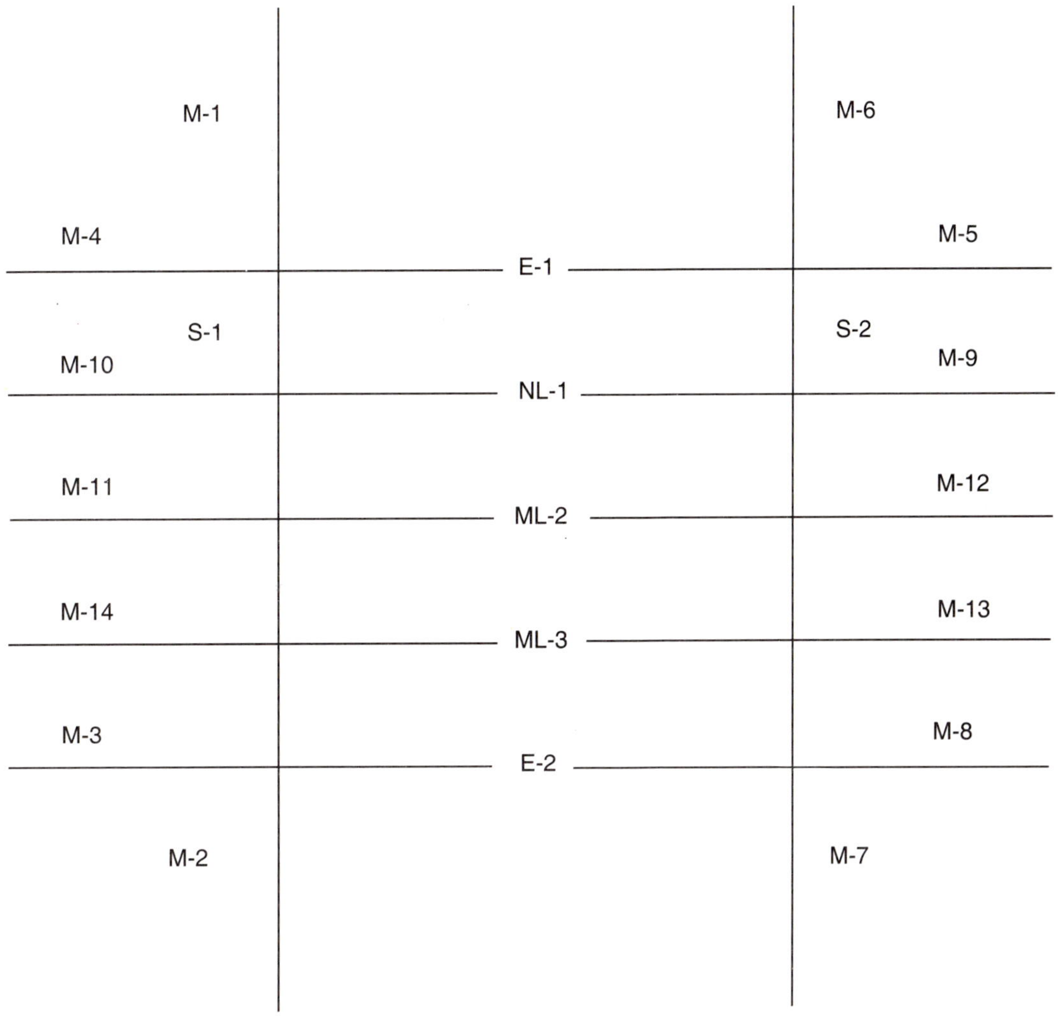

Figure 55. Open water floating production system configuration with internal system line details.

System Tuning

OBJECTIVE: The farmer will be able to properly tune a pole and floating system. The farmer will understand the necessity of tuning the system.

RATIONALE: For proper growth and net handling, tension within the system must be maintained. Tension on the net allows proper flow of seawater and therefore nutrients throughout the system. Tension within the system keeps nets from collapsing and keeps light even across nets within the system. Proper tuning allows for drying and conditioning in pole systems.

READING MATERIALS AND ACTIVITIES

1) Farmers will tune the pole and floating systems for the remainder of the course.

PRESENTATION: Tuning of the pole system consists of three things: pole integrity, net tension, and net height. Pole integrity is maintained by driving loose poles back into the bottom or replacing broken poles. Net tension is maintained by tightening loose ropes and equalizing tension on the net to maintain the shape of the net to maximize growing area. Adjusting net height is the tricky part. During neap tides, nets need to be attached higher to allow for proper exposure. During big tides nets can be lowered to reduce exposure. Cultivar and time of year determines the ideal net placement. Test nets can be tied in on a diagonal to determine the optimal height for latter nets to be installed.

Tuning of the floating system occurs at several points. A loose or sloppy system, loose production lines with a lot of play indicates tightening of anchors at the tie in points. All tuning of the floating system should be done around low water. Tensioning anchors will take most of the play out of a system. Net adjustments must be made at the net production line tie in points. Again nets need to be tied in as even as possible to assure even growth.

16. The mooring lines can now be tightened. Two healthy men in a boat can barely adequately tighten the system (be sure it is at low water!). A 25 HP outboard can pull about the same as two men. The boat can be used by attaching the mooring line to a Sampson Post (Figure 56) and, aiming the

Figure 56. Nori fishing boat "Sampson Posts" for "tuning" open water floating system.

boat directly towards the anchor, slowly increasing the engine throttle to full speed.

As the rope tightens, it can be squeezed together by a firm grasp of one hand at the point it enters the loop and tension is maintained. The other hand can then tie the sheet bend with a loop. This method of tightening requires the use of a tension anchor (Figure 41) to maintain some tension at low tides. The concrete block is tied onto the anchor line 20 - 30′ from the anchor, depending on how deep the water is. Placing a second boat with 2 men at the junction of the loop of the system line and mooring line, a 3/4" line is tied with a bowline adjacent to the loop, and the other end is wrapped around the Sampson Post on the boat. The engine is slowly throttled up to full power, and the mooring line is tightened and squeezed at the loop with a pair of vice grips by one worker, while the other worker ties the sheet bend with a loop. He then carefully unclamps the vice grips (be sure that there is a line on the vice grips and it is tied somewhere on the boat). The mooring line is not wrapped directly to the Sampson Post for safety concerns because of the extreme tension on the line. The system becomes very tight with this method.

17. See Figure 55 for proper system configuration. Attach M-9 to the first pair of loops on S-2 and deploy the anchor line and anchor. Always allow at least 10 ft. of extra line at the loop. Tie Middle line 1 (ML-1) to the other loop at the same point as M-9, and connect to S-1. Tie in 70-lb buoys at these junctions for flotation. Tie M-10 to the remaining loop at S-1 and deploy this anchor set. Attach the 2 other middle lines ML-2 and ML-3 in the same manner. These mooring lines can be tightened down, and observe the shape of the system to be sure it maintains the proper shape.
18. It is now time to attach yokes and mooring lines to every net row. Do not tighten these anchor lines down, and be sure that there is at least 10 ft extra mooring line through the loops of the yokes. There will be 25 yokes, mooring lines, and anchors at each end of the system. A mussel buoy is attached to yoke loop.
19. The production lines are now tied into the loops on the end and middle lines. The current or wind can sometimes be taken advantage of if the boat drifts in the proper direction to deploy the production lines. A 20 ft′ boat drifting sideways can deploy 3 production lines at one time with a three-person crew.
20. The last step is to tighten every anchor at the end of each net row. This is best done at low water and these lines must be very tight.

Net Handling and Installation

OBJECTIVES: The farmer will be able to assess the quality of nursery cultured fresh or frozen nets received. Thaw frozen nets in the appropriate fashion. Install nets in the appropriate manner. Assess frozen net acclimation post-introduction.

RATIONALE: A significant part of the success of a nori culture site is directly related to the quality of the post nursery-stage seeded nets it receives. The ability of the grower to determine the quality and acclimation the fresh or frozen nets is essential to the positive outcome of the culture grounds.

READING MATERIALS AND ACTIVITIES

1) Farmers will install fresh or frozen nets into pole and floating production systems.

PRESENTATION: Nori nets are cultivated in nursery systems until the thalli are approximately 2-3 cm long and removed from the nursery system and carefully transferred directly to the floating production system, remain in the fixed pole system, or are returned to shore to be frozen in bundles.

The nets are shipped frozen to the farmer for deployment. It is extremely important that the nets stay frozen to within two hours of being deployed. If the temperature of the nets rises above –7°C or the nets are thawed and refrozen, serious or complete failure will result. It is remarkable how fast a nori net can thaw, due to the lack of free water in the nets, which is key to successful preservation during the freezing process.

THAWING: Do not open the bag of nets until on site just prior to deployment. Place the bag of nets into the seawater, unopened, allowing the nets to adjust to the ambient water temperature. Five minutes is usually sufficient. This tempering is necessary to prevent potential damage to the young fronds. Should the storage bag be opened prior to the nets reaching ambient temperature, atmospheric fresh water will condense on the nori thalli resulting in plant stress or mortality.

When the bag is first opened, a pleasant odor will be emitted. If it is necessary to take a deep breath over the bag to detect this odor, this indicates potentially a good survival rate. A strong odor indicates potentially high mortality. Slowly add seawater to the bag to rehydrate the nori plants, five minutes is adequate. Observe the color of the nori at this time, it should resemble eggplant purple. A portion of the nori will exhibit a red-rust color indicating dead nori. If the red-rust color is very obvious or predominant, the nets should be suspect of being poor quality. All observations should be noted in a frozen net log and a report should be prepared upon installation as to the condition of nets received.

INSTALLATION: Place the bag of nets into the bow of the boat and maneuver the boat between the production lines or poles. All net end lines are tied together and a bundle of net end lines will be on top of the bag. Pull the bundle of net end lines out and examine the corner of the nets. All nori nets have a "2-strand corner" and a "3-strand corner" (Figure 57). Tie the 3-strand

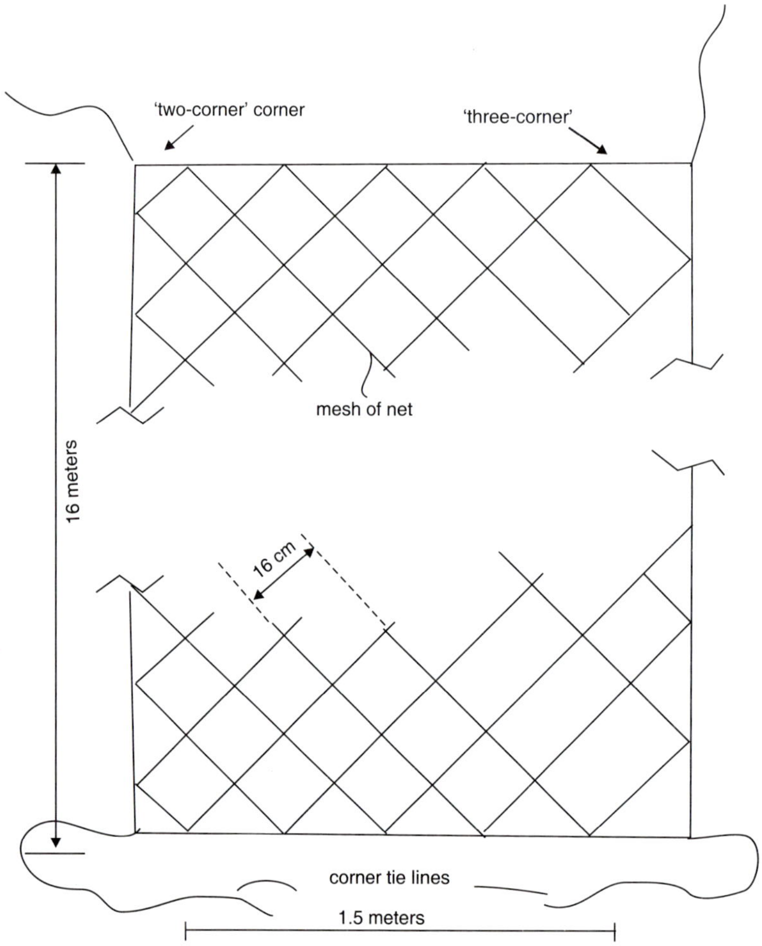

Figure 57. Schematic of nori net with the "2" and "3" corner delineation.

corner to the right side and the 2-strand corner to the left side (Figure 58).

This left side, right side distinction is important. When the nets are seeded, the plants on top of the net grow much faster and denser than on the bottom. It is critical to place the nets into the system with the top up. Having tied in one end of the net, move the boat backward toward the opposite end of the row, unraveling the nets out of the bag as the boat is moved. Grab the net ends as they appear out of the bag. Pull the bundle of net ends very tight to determine if the nets have rolled or twisted then, tie the end lines to the proper end support lines. At this point, one bundle of five nets will be tied into one row and now it is necessary to separate them. Untie one end of the net end lines and separate the top net, retie the remaining nets into the system. Pull the nets up over the bow letting the boat slide over the remaining nets. The top net should be laid into a basket as the boat is pulled along. When the other end of the net is reached, untie the net ends and separate the top net, retie the remaining nets into the system. Move the boat back to the original position and remove the next net in the same manner. Repeat this procedure until one net remains. Deploy single nets now in the boat securing the ends of the net front and back (remember to install the 3-strand corner on the right side).

It is time to tie in the sides of the nets. Move down the net to the net ties already in place. Use a sheet bend with a loop. Be sure to start the knot from the top of the net to the bottom so that the loop is on top of the net. The nets must be tied in taught and symmetrical.

Examine the nets for color and coverage. Some spots of rust red will be seen, the percent will vary as to the length of time the nets have been frozen. Nets frozen for longer periods have been seeded with higher spore or seed counts to allow for mortality and still produce a quality grow out net. The color and coverage should be documented in the frozen net log for the next 3 to five days. The net log should contain the following: net number and date frozen, approximate amount of rust color observed as a percent and any bare spots documented by size and location. The next day there will be an increase of bare spots, especially if the site has a strong current. It is not uncommon for 30% of good nets to be denuded within the first week.

These denuded areas will fill in by two methods. First, if a strand with no visible nori is

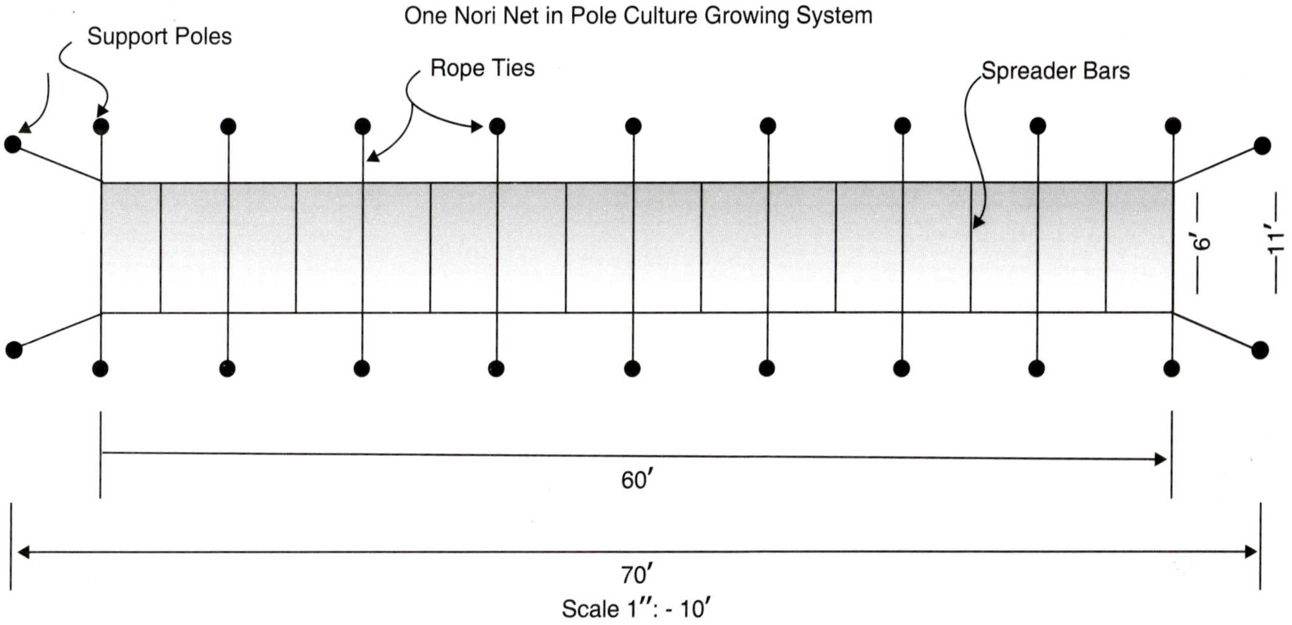

Figure 58. Schematic of nori net for a fixed pole production system.

examined under a microscope at 100×, small plants of 2 to 8 cells long (less than 0.2 mm) will be seen (Figure 59).

There will be from 10 to 100 of these plants per mm of net strand. Nori at this size freezes extremely well. If these plants are not present it is most likely that the net has been thawed for too long before placing in the ocean or re-frozen. Second, freezing stimulates the production of archeospores from nori on the nets. These spores are released from the margins of the existing plants and settle on the bare portions of the net strands, in effect re-seeding the nets. The spores can only settle on the net strands that have no nori on them. After a period of two weeks the bare spots should be covered with plants up to 1 cm long (Figure 60).

Daily Maintenance

OBJECTIVES: 1) The farmer will understand and be able to demonstrate normal care for the production nets and the production system. 2) The farmer will understand why certain daily practices for care occur. 3) The farmer will understand the ramifications of not performing the daily care practices.

RATIONALE: The practices for care of the nori nets and production systems have been developed over many years of nori culture. These

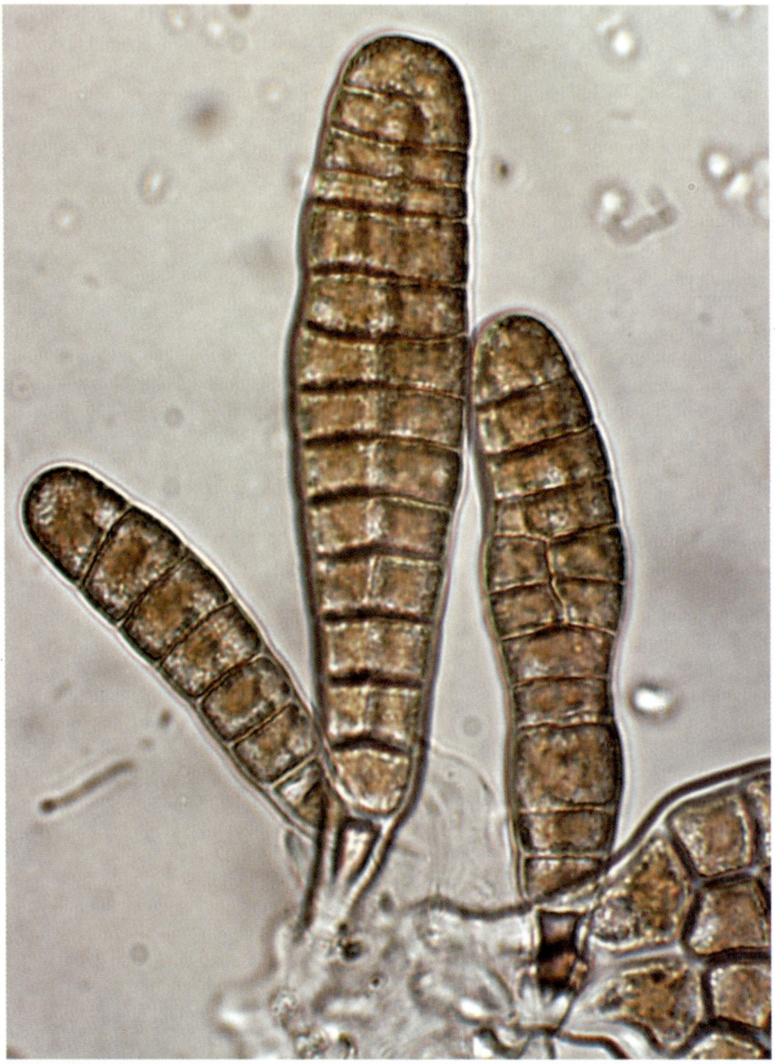

Figure 59. Emerging *Porphyra* haploid blades.

Figure 60. Haploid Blade ~ 1 cm in length (Courtesy: Professor Charles Yarish).

practices have proven effective in nori production and the lack of these practices can result in complete failure of the nets, system or both.

READING MATERIALS AND ACTIVITIES

1) Nets placed in both pole and floating production systems.
2) Spreader bars.
3) Have farmers install spreader bars and clean debris from nets.
4) Have farmers tune system.

PRESENTATION: Daily monitoring is required to achieve success. It is helpful to keep daily notes on the length and appearance of the nori. A careful examination will show daily changes in the nori's shape, length and color. Under ideal conditions the nori will grow in spurts of approximately 1cm/day. Usually a phytoplankton bloom occurs during these ideal conditions and the nori growth will slow as competition for nutrients increases. It should take approximately three weeks to grow from 2 to 15 cm at this time it is ready for the first harvest.

When the nori reaches 4-6 cm (Figure 61) long it is necessary to use 5′ PVC spreader bars.

At least six spreader bars are needed per net. These bars permit uniform fast growth all over the net. If not used the nori on the edges of the net where they are tied into the system will grow much faster. Nori that exceeds 20-24 cm before

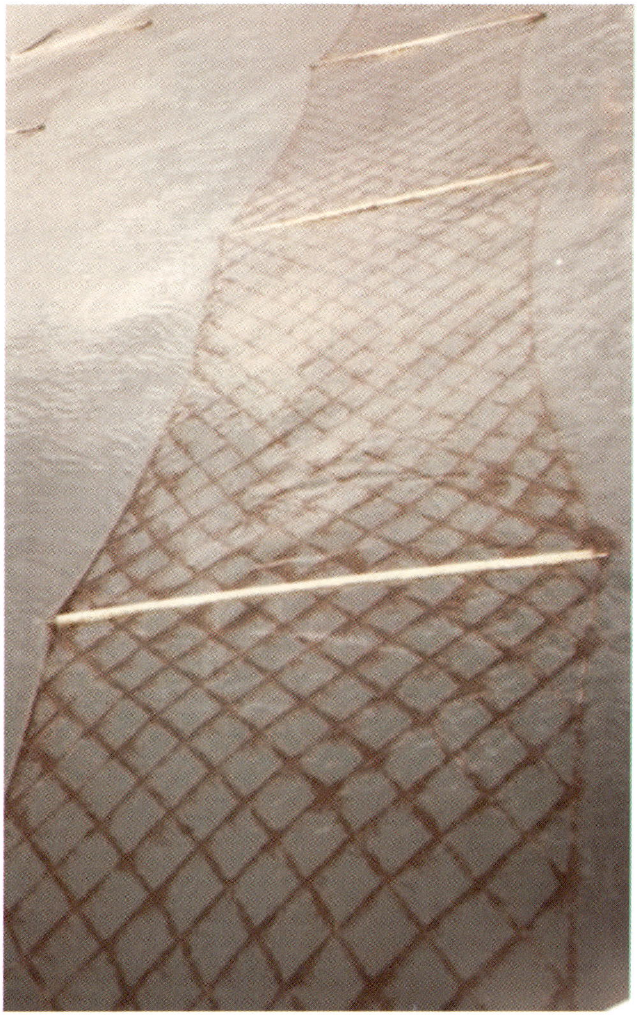

Figure 61. Nori net with young cultures

harvest becomes tough and because of its length shades smaller nori and out performs it. Disease can also be a problem in the higher growth areas. If the nori on the edges is 25 cm long while the nori in the middle is only 10 cm, a poor harvest will result and the net never will produce a proper quantity of nori.

All floating debris must be removed from the nori nets on a daily basis. This is extremely important on the day of the harvest. Extraneous materials harvested with the nori will decrease the quality even to the point of destroying the harvest. Floating seaweeds, plastics, and trees are serious concerns, especially during the spring tides. As debris sweeps into the nets it causes several problems. If the current drives the debris under the nets, the nets and nori thalli are raised above the water surface desiccating the nori. Seaweed rubs or abrades the nori off the nets over time with wave action. If debris floats over the top of the nets it shades the nori reducing its growth. Large mats of seaweed (Figure 62) and other debris can become entangled in the system decomposing and fouling the water. Large mats of seaweeds and other debris also become very hot on sunny days and the high temperatures will kill the nori plants.

Lowering the nori nets one-foot under the water surface allows the debris to float over the top and off the system and six-foot spaces or better between rows allow channels for the debris to flow through. A variety of booms and diversionary devices have been tried to prevent algae and other debris from entering the nori system with varying results. The most effective technique is to lay an extra row of nori nets on the sides of the nori system and to use them to collect debris. It can take 2-4 hours per day to clean a 100 net system. If a day is missed it may take 2-3 days to get the system clean again.

The nori nets must remain tight in the system. Loose nets cause distortion of the net resulting in unequal growing conditions. If the nets become very loose they may even fold over shading sections of the net and reducing growth. Tuning the system and nets is best done daily, ropes stretch and anchors move. The system will undergo more stress during the high running tides and require more adjustment during those lunar periods.

The nori thalli change color as a function of changing environment regimes, i.e. light and nutrient conditions. The best nori for making

Figure 62. Open water, floating nori production system with large mats of floating seaweeds (primarily *Ascophyllum* or rockweed) just prior to the daily system cleaning.

sheets has a rich purple, eggplant color and the darker the color the better the quality. It is helpful to obtain a Pantone Color Guide to be able to objectively monitor the plant's color.

Special Concerns

OBJECTIVE: The farmer will be able to properly assess, intervene and reassess a number of conditions common to nori farming.

RATIONALE: Farming is a hands-on activity where the changing environment determines how you proceed. Nori farming is a dynamic process and over time methods for the treatment of the most common conditions have been developed. One need not reinvent the wheel.

PRESENTATION: The successful nori farmer must carefully observe the growing nori on a daily basis. The crop will give indications of growing conditions by changing in size, shape, color, toughness and texture. Prompt identification of signs of poor growing conditions as shown by the plants will aid the farmer in attempting to ameliorate the situation. Generally, the sooner a problem is identified the easier it is to correct. The following section describes nori characterizations one might experience.

With increasing industrialization, many near shore waters have become polluted and unsuitable for cultivation. Historically treasured culture grounds have been abandoned, e.g. Tokyo Bay due to heavy metal and industrial pollution. Mitigation of the near shore water degradation is the development of offshore nori production system siting. Offshore cultivation experiences a different set of production challenges: exposure to storm and wave action disrupting the net anchoring system, reduced nutrient loading, and increased transportation fuel expenses.

Biotic challenges to either pole or floating cultures include: fungal disease (*Pythium porphyrae* – Red Rot; *Olpidopsis* – Chytrid Blight); bacterial disease (*Vibrio/Pseuomonas* – Green Spot; *Flavobacterium* – Anaaki "Hole Forming", *Leucothrix mucor* – Filament Disease) outbreaks (Fujita, 1990; Ishio et al., 1971; Kerwin et al., 1991, Levine, 2006, FAO, 2009), herbivore invasions and attachment of epiphytes that compete for light and nutrients (Oohusa, 1993a) (Figure 63).

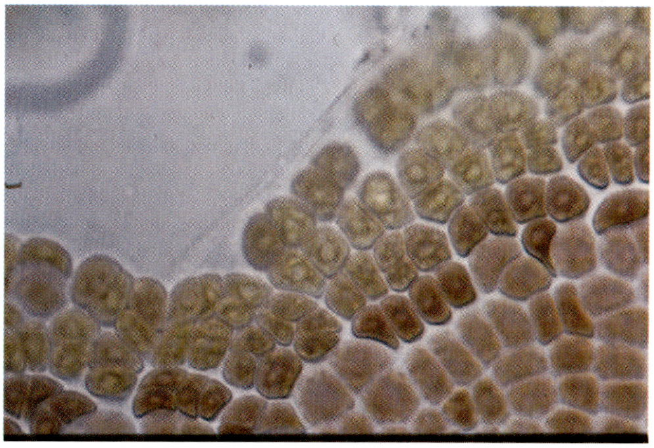

Figure 63. *Porphyra* disease, "Chytrid" found on production net.

Pale Color

Healthy, high quality nori has an eggplant purple color and should be observed at the base of the nori strand where numerous nori fronds overlap. One excellent technique is to cut a typical nori net strand and lay it on heavy matted drawing paper while immersed in water. Carefully remove the sheet from the water and lay on newspaper. Cover with wax paper and more newspaper and weight it down with a book to press the specimen flat. The sample will dry out over night and be preserved. It will maintain its color as long as it is kept in low humidity and away from light. Daily samples will give you a permanent record of color, growth, and morphology of the nori. Good color nori is 497-C in the Pantone Color Guide.

Nori can turn pale due to many causes; it can indicate a lack of nutrients, specifically nitrogen. This element is a key component of photosynthetic pigments such as phycobillins, which gives nori its purple color. In low energy areas 100 mg/L of nitrogen is required for proper nori growth. Frequent rainfall and subsequent runoff from terrestrial agriculture is required to maintain this level of nitrogen. In low energy areas additional space between nets will provide better water circulation.

If lack of nitrogen is suspected, ammonium nitrate can be safely added to the crop. Nori, < 7 cm long, will respond overnight to this treatment. If the crop is organically certified this treatment cannot be made. Successful applications have been made in several ways. For practical purposes it is not possible to over-fertilize and damage the nori. Laboratory experiments using 100,000 times more ammonium nitrate indicated better growth than untreated nori. The primary factor is economical. By placing 500 grams (about 1 lb.) or $.20 worth of chemical into 200 liters (50 gallons) sea water in a large tank and laying the nets in this solution for 5-10 minutes, 20-30 nets can be treated depending on how long the nori is and how dirty the nets are. Chemical costs are minimal and the only real expense is labor.

This fertilizer can be applied on top of the nets if the nets are on the surface of the water. A hand sprayer of 10-liter capacity is used placing 200 grams (0.5 lbs.) of ammonium nitrate in the sprayer and filling with sea water and the solution is sprayed from a boat moving up and down the rows of nets. Some success has resulted by treating 4 nets with each 10 liters. The nori has the ability to soak up the nutrients very quickly and store them for later use.

A third technique is to place ten kilograms of ammonium nitrate in a 6 mil plastic bag 50 cm long and 30 cm wide. Four slits 10 cm long are cut in the sides of the bag to allow seawater to enter. The bag is hung off a production line next to the nori net in a position so that the prevailing current sweeps the dissolved fertilizer into the net. Forty nets can be fertilized in this manner. The fertilizer dissolves in several days and then is replaced.

Pale nori also occurs when there is too much sunlight, which is common during June through August. The Japanese nori, *Porphyra yezoensis* is normally a winter crop thriving during the season of 10-hour days and 14-hour nights. Coastal areas experiencing extremely cold winters (0-6°C) will shift their culture season to the summer. Summer day's offer prolonged day lengths as a function of latitude. Just as with any plant cultured there are ideal amounts of light, dark and temperature. Nori can be darkened during the summer months by lowering the nets below the water surface. Nori exhibiting the finest color grew one meter above mean low water. At high tide the nets were approximately six meters below the surface.

Daily color changes will be noticed after a prolonged cloudy or foggy period. The best growing conditions for nori grown during the summer are several weeks of foggy weather, which frequently occurs in July. After September 15th the day length and solar intensity become satisfactory for nori and then a prolonged period of cloudiness results in pale color. After one day of sunny weather the nori will darken considerably.

Green Algae Infestation

Green algae, most commonly *Enteromorpha* sp. grows under similar conditions as nori. It will be found on bare net strands after 2-3 weeks deployment at any time during the growing season. If harvested with the nori it severely reduces the value of the harvested crop. The best defense against green algae is to have 100% nori coverage on the nets, which prevent it from attaching by the reduction of available substrate. When a few strands are found they can be pulled off by hand. If patches of green algae begin to grow on the nets the nets can be pulled from the system and given a long drying in the shade on a breezy afternoon. When salt crystals appear on the nori but the nori is still rubbery replace the nets back into the system. If this does not kill the green algae the nets can be frozen. They are dried in the shade and then placed in airtight bags and frozen at –18 °C for 8-10 days. This will kill the green algae but almost two weeks of production is lost. Nori nets with green algae can be treated with 0.2-0.5% citric acid solution. 1.5 kg (approximately 3 lbs.) of citric acid is dissolved in 200 gallons of seawater. The nets are placed in this solution for up to 10 minutes. Thirty nets can be treated with this same solution if they are not too dirty. Do not dry the nets before placing them in the solution. Always mix up the solution the day before use and place a few nori laden strands of net in the solution for 5, 10 and 15 minutes then place each strand in a different jar of seawater and keep cool. Examine the nori the next day for rust red color that indicates dead nori. This trial will help determine the length of time it is safe to soak the nori. An accurate pH meter is required.

Frayed Edges

Careful observation of the nori frond side edges shows rips or raggedness. Healthy nori edges are smooth with no serrations. The top of the nori frond will likely be ragged after the first harvest.

Abrasion from floating debris is the most common cause of frayed edges. If floating seagrass is in the area it can damage the nori because its edges are serrated and it wraps around the nori strand. As the current moves the seagrass blade along the strand, the nori is cut off or damaged. Daily cleaning of all floating objects from the system will limit this damage.

A sea state caused by strong winds can do extensive damage to nori fronds. Placing the nori site in protected waters is the only practical remedy.

Photooxidation, which can occur when periods of intense solar radiation occur along with reduced nutrients and/or low water temperatures, will cause ragged edges. Some cells on the margin of the nori are killed, and as adjacent cells continue to divide and grow, the area of dead cells cause a wrinkle to occur. If these environmental conditions are expected to continue, such as a drought period during the summer, then the best remedy is to lower the nets 3-5 meters (10-15 feet) below the surface. This is labor intensive and rather drastic, but these environmental conditions can cause complete failure of the nets for the season.

Crepeness

The edges of the nori fronds are wavy and curly, just as crepe paper appears. This condition indicates some physiological problem. The growth of the fronds will noticeably decrease and the nori tends to be tougher.

A possible cause is photooxidation or disease. This condition is most commonly found during the summer months and does not occur in nets fixed at 1 meter above mean low water or 6 meters below mean high water.

Frond Perforations

Perforations that are visible to the naked eye probably are caused by abrasion. If numerous holes exist, microscopic examination at 100× should be conducted. Holes of only 2-3 cells wide can indicate disease problems. A sample of nori net twine with fronds attached should be dried in paper towels or cheesecloth to remove most of the free water. The sample can then be placed in a zip-lock bag and shipped overnight for inspection. Be sure to remove all free water before shipping. The nori will survive for a surprisingly long time in a sealed zip-lock bad as long as it is kept dark, dry and cool. If the cells around the holes have a pale color or are totally clear, this may be an indication of disease. Prompt harvest must occur and removal of infected nets from the system must be conducted. Holes lined with a red ring visible to the naked eye may indicate red spot disease. Careful monitoring should be conducted. Several sightings of red spot in a whole system should not cause much alarm. However, if numerous sightings on one net are seen, immediately remove the net.

If holes larger than several cells wide occur and the cells around the margins of the holes have good brown-purple color, it is possible that sand has produced the holes. If the nori site is at the mouth of a large river and there have been heavy rains this "shot-hole" condition can occur. The only remedy is to move the nets to a different site or possibly freezing the nets until the environmental situation changes. Both remedies are rather extreme.

Additional Frond Color Changes

Pale edges near ends of frond: Male sex cells (spermatangia) appear as pale areas on the edges. Dark red spots in streaks in the frond: Female sex cells (zygotosporangia) are produced after the male sex cells. These sexual cells do not affect quality or production of the nori.

White edges of the fronds: This condition is usually caused by diatom infestation. If it is very widespread, the nori blades themselves will appear dull. Healthy nori shines. Special attention must be given to pole culture nets for diatom infestation when the low or neap tides occur around dawn and sunset. Pole culture nets are set low in the water so that they get about 4 hours of drying per 24-hour period. When the exposure times are at dawn and dusk the low solar intensity does not dry the nets and displace the diatoms. This tidal/solar period lasts between 5-6 days. Diatoms may be washed off with a pump with adequate pressure. The nets can be brought in to shore and hung in a shady area for 2-3 hours or until salt crystals form on the nori. The nori should remain rubbery or leathery. If it becomes so dry that the nori is hard and brittle the nori will die. Heavy diatom infestation will reduce the quality of the nori sheet. Such nori is called "hata-fun" nori or pigeon excrement nori, which is self-descriptive.

Low Growth Rates

Healthy nori in good growing conditions will show a daily increase in growth to the naked eye. Occasionally several days may go by without any noticeable change and then growth becomes obvious. A plankton bloom in the area may be the cause as the plankton uses up the available nutrients. If observable growth stops for 5 days or more it becomes economically detrimental. Careful observation of the nori fronds for signs of disease or excess plankton growth is essential. Growth stoppage during the summer can be due to a lack of nutrients, but the nori will almost always turn pale before it stops growing. Prolonged sunny conditions with low nutrients will adversely affect the growth and health of nori, to the point of killing it. Dark-colored nori that stops growing may be due to drops in water temperature.

Harvesting

OBJECTIVES: The farmer will be able to demonstrate the proper technique for harvesting from nori nets. The farmer will be able to list harvest options for the two main types of nori culture; pole and floating.

RATIONALE: The proper use of harvesting equipment and the timely harvest of nori plants is essential to the success of any nori growing operation.

READING MATERIALS AND ACTIVITIES

1) On land demonstration and hands on use of nori harvester.
2) On water demonstration and hands on use of nori harvester.

PRESENTATION: Quality is a function of but not limited to culture density, temperature, wind, rain, freezer net quality, pollution, plankton populations, disease, growth rates, storage conditions and harvest blade sharpness. Harvest quality however, will in great part be determined by two factors; net preparation and time of day. Immediately prior to harvest workers must move down the nets which are to be harvested and remove all foreign matter; seaweeds, pine needles, etc. If the net is harvested with foreign matter present the harvester will cut it along with the nori. Once mixed with the nori it is essentially impossible to remove and results in the harvest being unacceptable for processing. Time spent preparing the net for harvest is therefore essential to a successful harvest. Photosynthesis during the day changes the chemical components of the plant cells so that the quantity of sugars and amino acids at the end of the day are quite different than at dawn. Nori harvests need to occur first thing in the morning since this is when the proper levels of sugars and amino acids for sushi sheets are present. Nori harvested late in the day will produce a product with inferior aroma and taste, which is unacceptable. Exposure of the harvested nori to oil or gasoline from the harvester's gas engine or the boats motor will ruin that nori and if the contaminated nori is mixed with other harvested material it will ruin that as well.

The nori harvester has three blades which are 158 cm (70″) long, which is a little wider than the nori net (Figure 64).

A 5 HP Honda gas engine rotates the blades. The harvester is placed in a barge or boat and baskets with liners are placed beneath the harvester. The nets either in the floating system

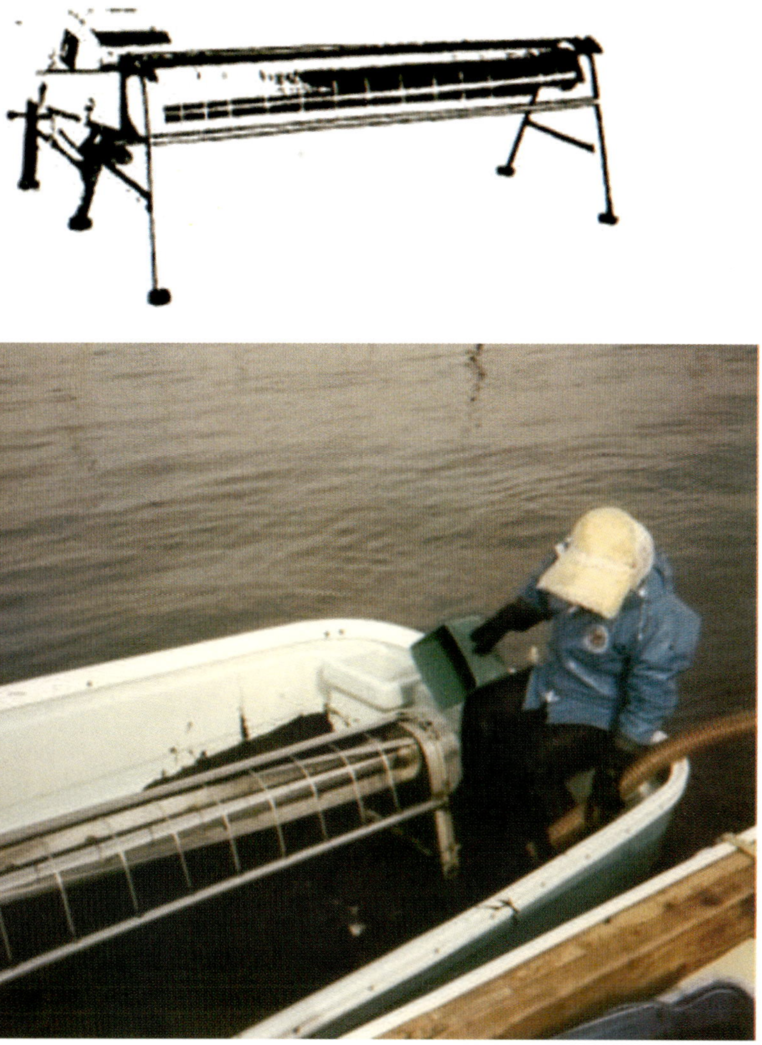

Figure 64. A) Nori harvester available in Japan and Korea (diagram). B) Japanese Harvester placed in two person floating maintenance and harvesting barge.

or pole system are dragged over the top of the harvester, the nets are still tied into their production system. The blades rotate in the direction that the boat is pulled under the nets cutting the nori off like a rotary grass mower. A movable panel on the top of the harvester controls the amount of blade exposed to the net. The least exposure leaves nori approximately 7.5 cm in length. The greatest amount of blade exposure strips the net bare and is used for the last harvest. In pole culture it is possible to harvest the nets without removing them from the poles at approximately one hour before mean low water. An alternative technique requires the removal of the nets into baskets in an adjacent boat or barge. The nets can then be pulled over the top of the harvester and placed back in baskets to be retied into the system. With practice, two laborers can untie, harvest and retie 40 nets in the two hours around mean low water. This method reduces the anxiety of completing this task within the very short window of the tidal cycle. Even if a problem occurs the nets can be tied onto the pole at a sub-optimal position and adjusted to the correct position during the next tide.

It takes two workers to harvest a net; one on

each side of the net to keep it stretched tight across the top of the harvester (Figure 65 a & b). If the net becomes slack it may wrap around the blades possibly cutting the net. One worker must be next to the engine to be able to stop the harvester at a moment's notice when a problem occurs.

Figure 65. Nori being harvested by two person teams.
A) floating system harvester, USA. B) Seto Inland Sea corporate nori farm duel net harvester, Japan.

The first harvest should occur when approximately 50% of the spaces between the meshes are filled with nori. The mesh size is approximately 15 cm (6″) between the net twine knots. This distance has been carefully determined to provide proper density and growing conditions for the nori. When a nori plant exceeds 15 cm it tends to become tougher and experience has shown that total production of the net is reduced. As the nori exceeds 15 cm the ends of the plants cover the next mesh causing competition between the nori plants and growth is affected. Some nori plants grow much faster than others and these are the plants first harvested. If the first harvest is delayed serious damage to the remaining crop can result. The nori should be cut back to about 7-8 cm (3″). It is much better to cut frequently than to allow the nori to become too long. The first harvest should result in about 10kg (22 lbs.) wet weight per net. The product is very soft and it can be difficult to make good sushi sheets from this harvest. Some Japanese farmers first-cut 20-30 nets and throw the first harvest away. The next week they harvest the same nets again getting a slightly tougher harvest product and first-cut another 20-30 nets to mix the harvests together. This method can result in a very high quality product (it is possible to freeze the first harvest and not waste any nori product, as you will see in the post-harvest handling section).

Under good growing conditions there should be 7-10 days between each harvest. By the 6th or 7th harvest the nori will have toughened and will no longer yield good product for sushi sheets. The harvester blades are then set to maximum exposure and the nets are striped clean. The nets are removed from the system and returned to the supplier. New seeded nets can be put in place if there is sufficient growing time left in the season.

Processing and Preservation

OBJECTIVES: The farmer will be able to demonstrate the proper handling and processing of post-harvest product for the preservation of the product quality. The farmer will understand the consequences of improper handling and preservation of post-harvest product.

RATIONALE: Nori plants are extremely friable and improper handling and preservation will make post harvest product valueless. The proper handling and preservation of post harvest product will preserve the quality and value of the post-harvest product for the grower.

PRESENTATION: The harvested crop is placed in netted baskets and allowed to be pressed dried during transport to processing area. Nori left in baskets begins to decompose immediately and the nori on the bottom of the basket becomes noticeably warmer with quality decreasing rapidly. The harvested nori product is brought in as soon as possible and must be processed into sheets or prepared for freezing within two hours post-harvest or quality decreases rapidly. If it is not possible to process within this time frame the nori can be stored in a seawater circulation tank with continuous water flow until it is time for processing.

The post-harvest nori product should never be left out in the sun, especially in the summer. It is best to place the harvest in insulated containers with covers. To cool the nori frozen gel packs should be used, placing 2-3 packs in the center of the baskets. If the farmer is more than two hours away from the processing facility then the harvest is frozen and shipped to the processing facility in this frozen state.

Nori sheet preparation starts with the harvested *Porphyra* rinsed in salt water to remove salt, silt, and diatoms; then minced into 2-6 mm pieces and then rinsed in fresh water. The slurry is pumped into a proportioning machine that accurately mixes the proper amount of seaweed with fresh water. The resulting mixture is pumped into large sheet making machine that essentially performs a papermaking process (Figure 66).

The product is run out on small mats to form a 19 by 21 cm wet cake where the water is squeezed out by two rows of sponge presses and then slowly dried in 35 – 45°C heat for several hours (Figure 67 a & b).

When dry, the sheets are peeled off the mat. This "nori" sheet weighs 3.0 g and is the traditional

Figure 66. Automated nori sheet making machine. Machines in come 5-8 sheet wide, some even double rows.

and standard Japanese product. The sheet product is packaged in moisture proof containers and may be stored for months in temperature and humidity controlled warehouses (Figure 68 a & b).

Following the drying process, the sheets are either sold at prefectoral auction or directly to the nori marketing companies. The dried nori is then redried prior to being roasted. The roasting of dried nori can be accompanied with soy or honey flavoring, a favorite with children (Figure 69 a & b).

Freezing the harvested nori must be preceded by cleaning and de-watering of the algal material. All foreign material must be removed from the nori and all free water must be removed or the freezing process will damage (most likely ruin) the harvest. Ten kilograms (22 lbs.) of nori, wet weight, is placed in a bag liner. Two nori filled liners are put in a centrifuge and spun until no water drips out of the drain. This will take 10-20 minutes depending on how quickly from the time of harvest this process is conducted. Each liner will then hold about one kilogram of de-watered nori. Since a good net will produce approximately 32 kg (70 lbs.) of product considerable time must be devoted to this de-watering process if 20 to 30 nets are harvested. Proper care must be given to the baskets of nori that have not been dried to prevent that nori from decomposing. Ideally the baskets should be placed in a walk-in cooler to prevent deterioration. Another option is the covered seawater tank with continuous flow of 20 gallons per minute. An agitator or paddle in the middle of the tank turning at a minimum of 15 rpm will prevent the nori from floating on the water surface and decomposing there. After de-watering the nori is placed in 6 mil plastic bags and tightly sealed. Each bag should hold no more than 2-kg (5 lbs.) so that rapid freezing can occur. The bags are placed in a freezer at –18 °C. The bags must have spaces maintained between them to allow rapid freezing to occur. These spaces can be easily

Figure 67. A) Nori sheets formed on plastic mats prepare for water removal by two rows of sponge presses; B) sheets travel up through drying box for two-hour drying process.

Figure 68. A) Finished sheets are checked for quality control, stacked and packaged. Wholesale cases will contain 36 to 40 packages of 100 sheets each. B) Some dried sheets are prepared directly for the retail markets.

Figure 69. A) Sheets are checked for quality and B) prepared for the automated, conveyor belt, radiant heat nori roaster.

produced between the bags with the use of small pieces of lath. The bags must be turned over after 24 hours to ensure complete freezing. The nori can be safely stored in this manner for several days and should be shipped to the processor as soon as it is frozen.

There are various alternatives to processing traditional nori sheets which include air-drying whole blades (these then may be used whole or milled into finer pieces), freeze-drying whole or chopped blades, making the raw wet blades into a product called "tsukudani-nori" (a soy sauce and nori paste), and using the raw nori fresh.

Factors Affecting Nori Quality

The production of high quality nori sheets is more of an art than a science. Quality is a function of but not limited to culture density, temperature, wind and rain, freezer net quality, pollution, plankton populations, disease, growth rates, storage conditions, and harvester blade sharpness. Additional factors affecting quality are the variety of nori grown, time of day of harvest, amount of sunlight, contamination of the harvest by other algae, small animals such as amphipods, shrimp, and other types of debris. Exposure of the harvested product to gasoline or oil from the gas engine on the harvester or the outboard motor will ruin the entire harvest.

The production and harvest of high quality nori thus results from a very personal involvement of day-to-day nurturing and management techniques that can only be learned by experience. One of the world's largest nori growers told the author that "It takes passion to produce good nori". Experience has found him to be right, and that there is a real danger to get caught up in strictly technological advances that make daily chores easier but distract the nori farmer from the daily nurturing required in the production of high-quality nori.

Organic Certification Requirements

The development of an organic nori farm represents a personal commitment by the farmer to follow a set of guiding principles and philosophies. These principles offer an alternative, healthier product which can be much harder to produce. Two farms, one in USA (Figure 70 a & b) and China, have earned this designation.

Organic certification requires strict record keeping. Each harvest has a lot number which describes the date and location of the harvest and is processed separately so that the processed sheets have a track record. This may result in a higher price for the sheets. If a nori farmer wishes to have his product organically certified, the farmer must follow certain guidelines in the nurturing of the crop, the harvest, and the care of the harvest. These are listed below.

- No chemicals can be used on the nori nets during grow-out.
- A designated barge must be used to harvest the nori to avoid contamination from outboard motor fuel or other sources of contamination.
- Fuel and lubricating oils must be kept away from the harvest. Remove the harvester engine from the barge before adding fuel to avoid spillage from contaminating the deck.
- A designated trailer or truck liner to haul the harvest from the barge to the storage/freezing area must be used.

Land-Based Nori Farming

Several attempts have been made to cultivate nori in land-based tank cultures, e.g. Hawaii and Maine, USA and Israel (Levine 2006). Natural seawater can be pumped from a submerged offshore inlet pipe into filtration system or artificial seawater can be utilized. Fresh, filtered, or artificial seawater utilized in controlled indoor systems avoid many of the external, nature driven environmental extremes, i.e. temperature, wind and waves, nutrient fluctuations, epiphytes, herbivory, disease, culture contamination, debris entanglement, etc. Maintenance of ideal growing conditions is facilitated in tanks, allowing easier environmental manipulation resulting in increased yields and thalli quality. Increased yields and nori quality must be achieved to offset, in part, the higher capital expenditures associated with land-based operations.

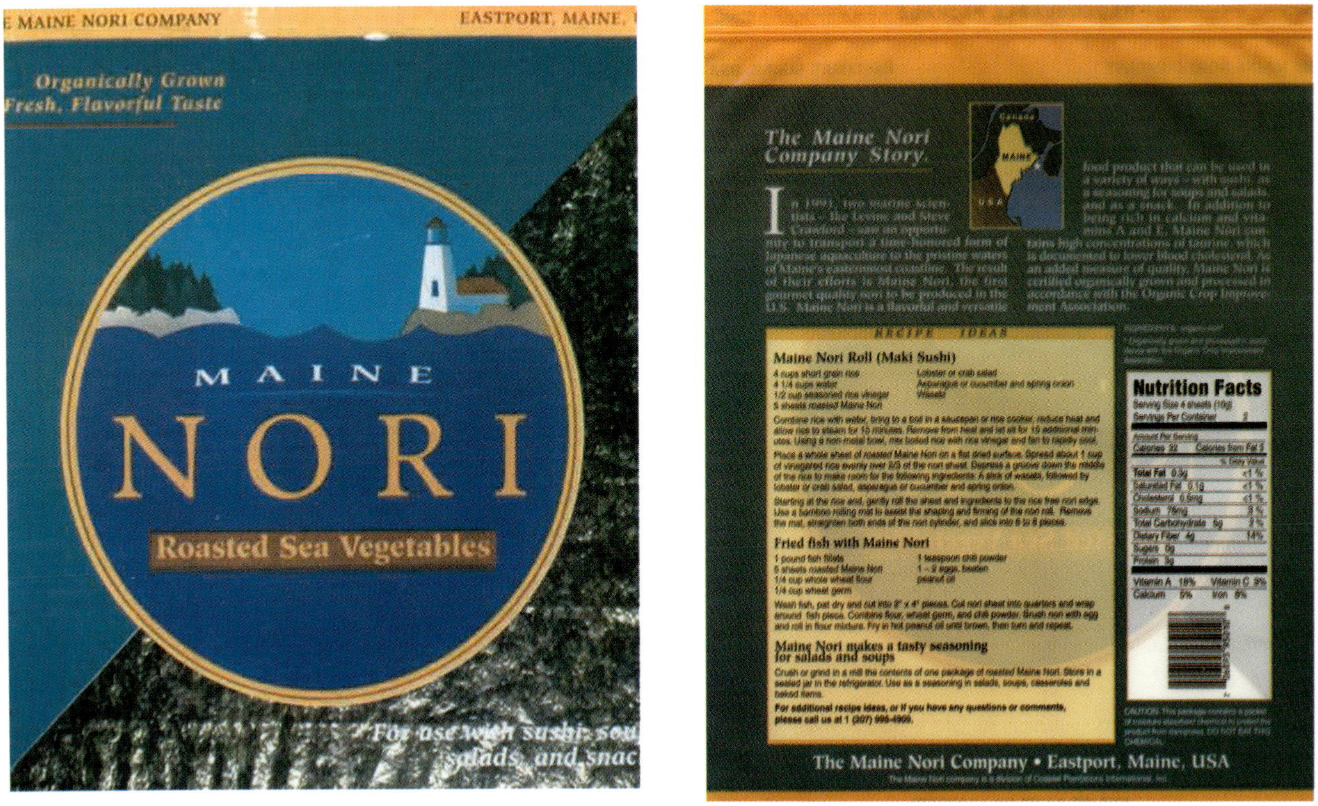

Figure 70. A) The packaging of the world's first organically farmed and processed nori. A) Front panel includes the organic certification from OCIA – Organic Certification International Association. B) The rear panel offers *Porphyra's* nutritional composition and a recipe for its use.

Land-based *Porphyra* cultivation systems experience additional technical challenges that must be resolved before tank cultivation can be considered a commercial option. Seeded nets could simply be placed in the tanks for grow out, but this would be a waste of space. A better use of the tank's volume is the production of suspension cultures. Blades in these cultures are not attached to a substrate and are suspended freely in the water. Because the tanks represent a three-dimensional volume (nets are essentially two dimensional) less area is required for a commercial land-based farm than for an ocean-based farm. As land-based tank production values have not overcome the additional capital and operation costs of this farming strategy, it remains a unique and situational opportunity at best.

6 Conclusion

Seaweed farming in general and *Porphyra* farming in particular remains a viable growth industry for food, feeds, fuels, and biotechnology. *Porphyra* farming, as illustrated in this volume, represents the science, engineering, and art of this centuries old profession. The choice of cultivation formats; fixed pole, floating, or semi-floating systems is a function of local culture conditions, access to culture grounds, available system assembly materials, and personal farming preferences.

The generosity of spirit and technology offered by the Japanese nori fishermen, nori fishing cooperatives, prefectoral extension agents, university-based academics, nori processing and marketing companies, and national nori advocacy organizations was both overwhelming and essential to our success in bringing nori farming to North America. Equal hospitality was afforded by the Chinese nori farming villages and the scientists of Academica Sinica, Qindao, China. The ability to spread nori farming to new parts of the globe rests in the hands of experienced professionals willing to share their knowledge and experience. This generosity of spirit is presently alive and well.

The Indian nori farming initiative represents the potential for the next great cultivation area. The passion for *Porphyra* farming is infectious and represents an effective socio-economic engine for the development of coastal villages. Any individual or group who is interested in taking up *Porphyra* farming should contact their local aquaculture extension agent or the authors for more information. Like any farming effort, experience goes a long way in improving yields and quality. The only way to get started is to get one's feet wet.

Literature Cited

Arasaki, S. (1982). Advances in phycological research in fishery sciences in Japan. *Jpn. J. Phycol.*, **30:** 390-400. (In Japanese).

Chopin, T., Yarish, C., Wilkes, R., Belyea, E., Lu, S. and Mathieson, A. (1999). Developing *Porphyra*/salmon integrated aquaculture for bioremediation and diversification of the aquaculture. *J. Applied Phycology* **11:** 463-472.

Egan, B. (1990). Nori farming star of the '90's? *The Fish Farm News* **3(7):** 25-27.

Fedlmann, J. (1979). Sur une nouvelle espece de *Porphyra* (*P. drachii* J. Feldm.) et remarques sur l'évolution des plastes chex les Rhodophycées (Bangiophycidées et Acrochaetiales). Vie et Milieu, **28-29:** 425-435.

Food and Agriculture Organization of the United Nations. 2009. Review of the state of world aquaculture, Inland Water Resources and Aquaculture Services. FAO Fisheries Circular, Available at: www.fao.org

Fujita, Y. (1990). Diseases of cultivated *Porphyra* in Japan. *In:* Akatsuka I, ed. Introduction to applied phycology. S.P.B. Academic Publishing, The Hague, pp. 177-190.

Hafting, J. (1998). Strategies for growth management of *Porphyra yezoensis* Ueda blades in suspension cultures: a step towards land based mariculture. Ph.D. Thesis, University of British Columbia, 46-47.

Hardy, R.W. (2000). Urban legends and fish nutrition. *Aq. Mag.* **26(6):** 47-50

Jensen, A. (1993). Present and future needs for algae and algal products. *Hydrobiologia* **260/261:** 15-23.

Ishio, S., Yano, T. and Nakagawa, H. (1971). Cancerous diseases of *Porphyra tenera* and its causes. *In:* Nisizawa, K, ed. *Proc. 7th Int. Seaweed Symp.*, University of Tokyo Press.

Kerwin, J.L., Johnson, L.M., Whisler, H.C. and Tuininga, A.R. (1991). Infection and morphogenesis of *Pythium marinum* in species of *Porphyra* and other red algae. *Can. J. Bot.*, **70:** 1017-1024.

Levine, I.A. (1997a). North American *Porphyra* cultivation - from molecules to markets. Second Asia-Pacific Marine Biotechnology Conference and Third Asia-Pacific Conference on Algal Biotechnology, May 7-10, 1997, Phuket Thailand. Bangkok, Thailand: National Center for Genetic Engineering and Biotechnology, 53 pp.

Levine, I.A. (1997b). US *Nori* farming: an emerging industry. *Phycological Newsletter*, **34(2):** 16-18.

Levine, I.A. and Crawford, S. (1997). North American *Porphyra* cultivation - marine agronomy, integrated polyculture and biotechnology. *Phycologia* **36(4):** 60.

Levine, I.A. (1998). Commercial cultivation of *Porphyra* (*nori*) in the United States. *World Aquaculture*, **29(4):** 37-48.

Levine, I.A. and Cheney, D. (1998). North American *Porphyra* cultivation - from molecules to markets. In: Le Gal Y, Halvorson HO, eds. New Developments in Marine Biotechnologies. New York, USA: Plenum Press, **(30):** 141-144.

Levine, I.A. (1999). Marine agronomy in the Maritimes: reality or fantasy? Bull. Aquaculture Assoc.-Canada, **99(1):** 31-33.

Levine, I. (2006). Case study on *Porphyra* cultivation in Maine, USA and New Brunswick, Canada. In: Aquaculture Compendium, Online at www.cabicompendium.org/ac. Wallingford, UK: CAB International. pp. 1-13.

Lindstrom, S.C. and Cole, K.M. (1992). A revision of the species of *Porphyra* (Rhodophyta, Bangiales) occurring in British Columbia and adjacent waters, *Can. J . Bot.*, **70:** 2066-2075.

McDermid, K.J. and Stuercke, B. (2003). Nutritional composition of edible Hawaiian seaweeds. *J. Appl. Phycol.*, **15(6):** 513-524.

Mitman, G.G. and Van der Meer, J.P. (1994). Meiosis, blade development and sex determination in *Porphyra purpurea* (Rhodophyta). *J. Phycol.* **30:** 147-159.

Miura, A. (1975). *Porphyra* cultivation in Japan. *In:* Tokida, J. and Hirose, H. Eds. *Advance of Phycology in Japan* Dr. W. Junk, The Hague, 273-304.

Miyata, M. and Kikuchi, N. (1997). Taxonomic study of *Bangia* and *Porphyra* (Bangiaceae, Rhodophyta) from Boso Peninsula, Japan. *In:* Miyata, M. and Notoya, M. (Eds.), Present and future on biology of Porphyra, *Nat. Hist. Res.*, Special Issue, **3:** 19-46.

Mumford, T.F. Jr. and Cole, K. (1977). Chromosome numbers for fifteen species in the genus *Porphyra* (Bangiales, Rhodophyta) from the west coast of North America. *Phycologia* **16:** 373-377.

Mumford, T.F. (1990). Nori cultivation in North America: growth of the industry. *Hydrobiologia* **204/205:** 89-98.

Nelson, W.A. (1993). Epiphytic species of *Porphyra* (Bangiales, Rhodophyta) from New Zealand. *Botanica Marina*, **36:** 525-534.

Noda, H. (1993). Health benefits and nutritional properties of nori. *J. App. Phycol.*, **5:** 255-258.

Notoya, M. (1997). Diversity of life history in the genus *Porphyra In:* Miyata, M. and Notoya, M. Eds, Present and future on biology of *Porphyra. Nat. Hist. Res.* Special Issue **3:** 47-56.

Okazaki, A. (1971). Seaweeds and their uses in Japan, Tokai University Press, Tokyo, pp. 165.

Oohusa, T. (1993a). Recent trends in nori products and markets in Asia. *J. App. Phycol.* **5:** 155-159.

Oohusa, T. (1993b). The cultivation of *Porphyra* "Nori". *In:* Ohno, M. and Critchely, A.T. eds. Seaweed cultivation and marine ranching, Yokosuka, Japan, pp. 57-73.

Piriz, M.L. (1981). A new species and a new record of *Porphyra* (Bangiales, Rhodophyta) from Argentina. *Bot. Mar.* **24:** 599-602.

Sahoo, D.B. (2000). Farming the ocean: Seaweeds Cultivation and Utilization. Aravali Books International, New Delhi, pp. 44.

Sahoo, D.B. (2005). Biology, Cultivation and Economics of *Porphyra* spp. In Aquaculture compendium 2005 Edition. CAB International. Wallingford UK. 1-40

Sahoo, D.B. and Nivedita, Debasish, (2001). Seaweeds of Indian coast. APH Publishing Corporation, New Delhi, pp. 283.

Sahoo, D.B., Tang, X. and Yarish, C. (2002). *Porphyra* – the economic seaweed as a new experimental system. *Curr. Sci.* **83(11):** 1313-1316.

Sahoo, D.B. and Yarish, C. (2005). Mariculture of seaweeds. *In:* Anderson, R. eds. Algal Culture Techniques. Academic Press, USA, 219-238.

Sahoo, D.B., Baweja, P. and Kushwah, N. (2006). Developmental studies in *Porphyra* vietnamensis – a high temperature resistant species from Indian coast. *J. Appl. Phycol.* **18:** 279-286.

Tang, X.R. and Fei, X.G. 1999. Artificial cultures from conchocelis to conchosporelings of *Porphyra katadi* var. *hemiphylla*, *Ocean Limnol. Sci.* **30:** 180-185.

Tseng, C.K. and Fei, X.G. (1987). Microalgal commercialization in the Orient. Hydrobiologia, **151/152:** 167-172.

Tsujii, K., Kchikawa, T., Matusura, Y. and Kawamura, M. (1983). Hypercholestrolemic effect of taurocyamine or taurine on the cholesterol metabolism in white rats. Sulphur Amino acids, **6:** 239-248.

West, J.A. and Hommersand,, M.H. (1981). Rhodophyta: life histories. *In:* Lobban, C.S. and Wynne, M.J. Eds. The Biology of Seaweeds. Blackwell. Oxford, 133-193.

Yoshida, T. (1997). Japanese marine algae: New combinations, new names and new species. *Phycological Research* **45:** 163-167.

Yoshida, T., Notoya, M., Kikuchi, N. and Miyata, M. (1997). Catalogue of species of *Porphyra* in the world, with special reference to the type locality and bibliography. *Nat. His. Res.*, Special Issue, **3:** 5-18.

Personal communication; Operations Director of Takaokaya USA, one of the world's five largest nori marketing companies. November 4, 2009.

Glossary of Terms

Algae: Photosynthetic eukaryotic organisms that generally reproduce by unicellular sex organs. Inhabit marine, brackish, and fresh waters. Additionally found in or on soil, rocks, glaciers, volcanic steam vents, etc.

Algal Ecology: The branch of algae, which deals with the interrelationships between living things and the relationships between organisms and their environment.

Antioxidant: A compound or substance, which actively binds to (renders inactive) or inhibits and/or prevents the formation of free radical compounds.

Aquaculture: farming animals or plants in the water (marine or freshwater)

Aquatic Effluents: The outflow from municipal solid waste treatment facilities or industrial waste discharges.

Archeospore: (monospores) asexual reproductive spore developed by vegetative cells along the margin of emerging blade phase.

Bioremediation: The use of living material to degrade environmental hazards or polluting substances.

Blade Phase: macroscopic, haploid, vegetative "leaf-like" stage of the *Porphyra* life history. Blades phase can be either monoecious or dioecious.

Conchocelis phase: The diploid, microscopic, filamentous, calcium carbonate boring, sporophytic phase of *Porphyra* life history.

Conchosporangia: "fat filaments" which differ from vegetative conchocelis filaments by a change in size, shape, color, cell morphology, and branching pattern.

Conchospores: diploid cell, developed in conchosporangium located on "fat filaments" of the microscopic conchocelis phase of *Porphyra* life history. The cell undergoes meiosis immediately upon release and germinates to form the haploid, foliose, blade phase.

Cultivar: A contraction of "cultivated variety". It refers to a plant type within a species that is distinguished by one or more characters; horticulturally, such plants are of considerable economic importance.

Epiphyte: An organism that grows upon a plant but is not parasitic upon it.

Heteromorphic: different looking phases of the life history. *Porphyra* has both a macroscopic blade and microscopic filamentous phase.

Ikada: the Japanese nori fishery term to indicate the nursery blade culturing phase. Key to this phase is the daily emersion of the nets for drying the nets and reducing contaminants.

Mariculture: farming animals or plants in the ocean.

Monoculture Systems: The cultivation system utilizing a single genus and species as compared to polyculture.

Multitrophic aquaculture: an expansion of the term polyculture to indicate different trophic levels, e.g. producers, primary consumers, secondary consumers, present in the same system supporting each with nutrient secretion or absorption.

Nori Physiology: The branch of algae (nori) science, which deals with the various processes, and activities which occur within a living organism or one of its parts; the study of function.

Nutraceuticals: Those foods and/or supplements that are thought to enhance health or decrease the onset or severity of illness or disease.

Phycobilin Pigments: A group of water-soluble, protein-linked pigments found in blue-green bacteria and the red algae.

Phycoremediation: The development of bioremediation using algae.

Polyculture: The multispecies cultivation in aquatic agronomic systems.

***Porphyra* (nori):** A member of the phylum Rhodophyta (red algae): Class Bangiales. A primitive (up to 1.2×10^9 years old) genus located in estuarine and marine waters worldwide. One to two cells thick, the inter- and subtidal alga resembles red leaf lettuce in appearance.

Recruitment: the release of reproductive cells and the successful attachment to a substratum for their subsequent germination and growth.

Saku: the Japanese nori fishery term to indicate the spacing in a production array to hold one net.

Seaweeds: macroscopic forms of algae.

Zygotospores: diploid spores produced via mitotic divisions forming dense assemblages of cells. The cell wall ruptures and the zygotospores are released which germinate into the conchocelis stage of the *Porphyra* life history.

Index